8/83

Dynamic Analysis of Offshore Structures

Dynamic Analysis of Offshore Structures

C.A. BREBBIA, DIPL.ING., PH.D.

Senior Lecturer in Civil Engineering,
University of Southampton, UK

S. WALKER, M.A.(CANTAB.), M.SC.

Consulting Engineer,
Structural Dynamics Ltd, Southampton, UK

NEWNES–BUTTERWORTHS
LONDON   BOSTON
Sydney   Wellington   Durban   Toronto

The Butterworth Group

United Kingdom **Butterworth & Co. (Publishers) Ltd**
London: 88 Kingsway, WC2B 6AB

Australia **Butterworths Pty Ltd**
Sydney: 586 Pacific Highway, Chatswood, NSW 2067
Also at Melbourne, Brisbane, Adelaide and Perth

Canada **Butterworth & Co. (Canada) Ltd**
Toronto: 2265 Midland Avenue, Scarborough,
Ontario M1P 4S1

New Zealand **Butterworths of New Zealand Ltd**
Wellington: T & W Young Building,
77–85 Customhouse Quay, 1, CPO Box 472

South Africa **Butterworth & Co. (South Africa) (Pty) Ltd**
Durban: 152–154 Gale Street

USA **Butterworth (Publishers) Inc**
Boston: 10 Tower Office Park, Woburn, Mass. 01801

First published 1979

© Butterworth & Co. (Publishers) Ltd, 1979

British Library Cataloguing in Publication Data

Brebbia, Carlos Alberto
   Dynamic analysis of offshore structures.
   1. Offshore structures – Dynamics
   I. Title   II. Walker, Stephen
   627′.98   TCI650   79–40457

   ISBN 0–408–00393–6

Typeset by The Macmillan Co. of India Ltd., Bangalore
Printed in England by Billing & Sons Ltd., Guildford, London and Worcester

Preface

The discovery of large deposits of oil and gas in deep-water regions has resulted in the construction of large production and drilling platforms, which often have to withstand severe environmental conditions in inhospitable areas. Often the natural period of vibration of these structures is comparable with the typical period of variation of the environmental forces, making a dynamic spectral analysis necessary.

This book starts by explaining the fundamentals of probabilistic processes, develops the theory necessary for the description and analysis of sea states and describes the random-vibration approach to structural response. Chapter 3 explains the essential hydrodynamics of water waves, describes wave forecasting, and defines the statistical parameters associated with sea state description. Chapter 4 deals with the calculation of wave forces on slender members (such as those used in the construction of steel lattice-type structures) using Morison's equation, and Chapter 5 describes extensively the use of diffraction theory in the calculation of wave forces on large-diameter bodies such as those found on concrete gravity-type structures. In Chapter 6, the study of environmental forces is completed by describing the effect of currents and winds.

Chapter 7 is an introduction to the theory of vibration, including a description of the spectral approach, and Chapter 8 describes the extension of this theory to multi-degree-of-freedom structures, as well as giving a brief introduction to the method of matrix analysis of structural response. Finally, in Chapter 9 some case studies are reported, and the problems of fatigue and soil–structure interaction are discussed.

The text is suitable for one-year courses in offshore structure analysis at postgraduate or final-year undergraduate level, and it should also prove useful for study by practising structural, civil and maritime engineers.

Financial support from the Science Research Council and Southampton University is gratefully acknowledged. We would also like to thank Mr M. A. McSweeney of Southampton University for dealing with the administrative problems arising from the production of this book, and Mrs Audrey Lampard for typing the manuscript in such a faultless way.

C. A. B.
S. W.

Contents

CONTENTS

1 Introduction

1.1 INTRODUCTORY REMARKS

The construction of offshore structures in ever-increasing water depths has necessitated the reassessment of their methods of design. Since the 1940s the tendency to build fixed offshore structures in deeper water has progressed steadily (Figure 1.1). Exceptional examples are the steel fixed platform for the Santa Barbara Channel erected in 260 m of water and the one designed for 300 m depths in the Gulf of Mexico. With exploration in water depths of around 1000 m, these

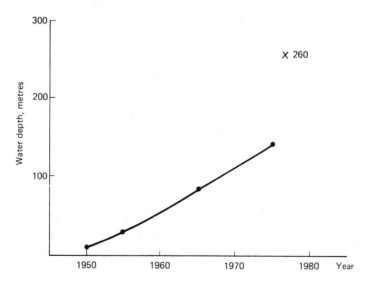

Figure 1.1 Maximum water depth for offshore structures continues to increase

1

new designs may indicate an acceleration in the rate of growth of the size of offshore platforms.

It is important both to determine how far our present knowledge can be extrapolated in order to analyse deep-water structures, and to understand the most up-to-date methods of analysis, which can often produce better estimates of the structural response.

Special care should be taken to understand the shortcomings of the different analytical steps. Current design processes start by assuming estimates of wind and wave climates and other forces (Figure 1.2), which may be very approximate. These estimates then need to be transformed into loads acting on the structures, and in this way new sources of uncertainty are introduced. The structural response produced by such loads can nowadays be very accurately determined, for linear systems, by using computational techniques. The main uncertainties occur when trying to estimate the soil properties and the

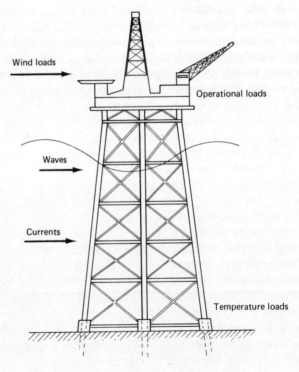

Figure 1.2   Loads acting on an offshore structure

fatigue life of the structure. Design guidelines on these two problems are scanty and unreliable, owing to the lack of experimental data and the need for more adequate analytical techniques. In addition to all these uncertainties one should take into consideration the possibility of substantial construction and material variations. This book aims to provide the analyst with a state-of-the-art appraisal of the analysis of offshore structures, pointing out the major sources of uncertainty in the design process.

1.2 SEA STATES

The response of offshore structures to wave loading is of fundamental importance in the analysis. Waves account for most of the structural loading and, because they are time dependent, produce dynamic effects tending to increase the stresses and damage the long-term behaviour of the system.

Wave forecasting techniques allow us to determine the sea states that will occur during the life of the structure, and in particular the worst case, usually the sea state corresponding to a storm with a return period of 100 years. While an understanding of the worst possible storm is necessary to determine the maximum stresses and displacements in the structure, knowledge of *all* the sea states allows us to calculate long-term statistics. These long-term statistics are usually more important, since the maximum stresses occur only during a very short period in the life of the platform.

The determination of sea states will be discussed in detail in Chapter 3, but Figure 1.3 summarises the two main techniques for defining them. In both cases the minimum information required is the wind distribution or 'wind rose'. This information, which is usually presented in terms of the Beaufort scale, can be transformed into a wave scatter diagram by relating the wind and its duration to the basic parameters of a sea state, e.g. significant wave height and mean zero crossing period. This conversion can be carried out using graphs such as those due to Darbyshire, Draper and others (see Chapter 3), which take into account fetch and duration of the wind. From the wave scatter diagram one can pass to either a wave height exceedance diagram or several wave spectral density curves.

The wave height exceedance diagram is obtained by plotting height

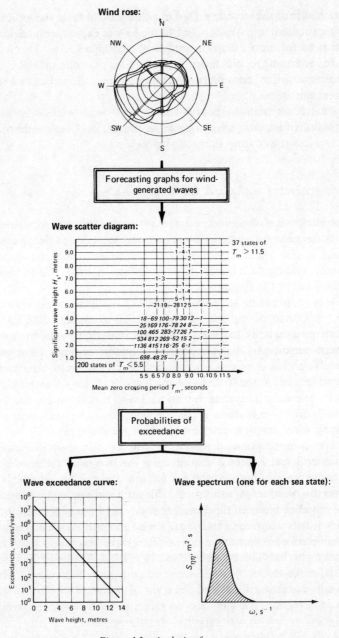

Wind rose:

Forecasting graphs for wind-generated waves

Wave scatter diagram:

37 states of $T_m > 11.5$

200 states of $T_m < 5.5$

Probabilities of exceedance

Wave exceedance curve:

Wave spectrum (one for each sea state):

Figure 1.3   Analysis of sea states

versus number of waves on a semi-log graph. The results can then be approximated by a straight line. This diagram can be built up for all waves or for waves from a particular direction. (Directionality of waves is important for fatigue calculations, as it can give reduced estimates of long-term damage.) The results can be extrapolated to determine the maximum wave for a given return period. Plots of wave height versus period can also be obtained from the wave scatter diagram, and are approximately on a straight line when plotted on a semi-log graph; from these results the maximum wave for a given return period can be deduced.

The full predictive process requires the use of forecasting graphs for wind-generated waves and the assumption that the probability of exceedance can be expressed mathematically (usually as a Rayleigh-type probability function).

In the spectral approach, exceedance diagrams are not used. Instead each of the sea states shown in the wave scatter diagram is transformed into a spectrum. These spectra are then applied to study the response of the system in a probabilistic manner. The spectra are usually applicable to fully developed seas, although some of them are not. A source of uncertainty here is the particular form of the spectrum used (see Chapter 3). More important than the differences in spectral maxima are the differences in the higher-order frequency part or 'tail' of the spectrum, i.e. those affecting more the structural response. In addition to this, measured spectra are not unimodal, but have a series of secondary peaks at higher frequencies, usually multiples of the maximum peak frequency. These peaks have an influence on the response, but they are neglected and a smooth analytical expression is usually used instead.

The spectral approach can be used to determine the maximum response by working with the spectrum of the centenary wave. Once the probabilistic analysis is carried out and the standard deviation of the response found, one can multiply it by a constant value $\pm \lambda$ (usually taken between 3 and 4 for Gaussian processes), which gives the maximum response within a certain probability. Another possibility is to take into account the record duration, i.e. duration of the storm, and then compute the expected response using formulae such as those presented in Chapter 2.

1.3 WAVE FORCES

The determination of the forces exerted by waves on structures is a very complex task, even when considering slender members. Waves can be represented analytically using different theories, but they may produce additional effects (such as slamming and slapping on the structural members) that defy simple analytical treatment. The forces may also be the result of an interactive process between the structure and the waves, with the former modifying the kinematics of the waves.

From the offshore structure designer's point of view it is important to differentiate between the various force regimes resulting from the interaction between the structure and the waves. They can be summarised as follows:

- for $d/\lambda > 1$ (d = diameter or characteristic dimension and λ = wavelength), conditions approximate to pure reflection;
- for $d/\lambda > 0.2$, diffraction is increasingly important;
- for $d/\lambda < 0.2$, diffraction is negligible;
- for $d/w_o > 0.2$ (w_o = orbit width parameter equal to wavelength in deep water), inertia is dominant;
- for $d/w_o < 0.2$, drag becomes more important.

The determination of the forces resulting in any of these regimes can be formulated in two basic steps: first, computation of the kinematic flow field starting with the water elevations, using a wave theory; second, determination of hydrodynamic forces applying Morison's equation or diffraction theory.

Morison *et al.* (see Chapter 4) developed a formula to represent the total force per unit length on a slender cylinder. The formula takes into account the inertia and the drag components, i.e.

$$F(t) = C_I \dot{v} + C_D v|v| \tag{1.1}$$

The force is in the direction of wave advance, and the water particle velocity v and acceleration $\dot{v}$ are evaluated at the cylinder axis. C_I is a constant due to inertia consisting of two terms, one due to the 'hydrodynamic' mass contribution and the other to the variation of the pressure gradient within the accelerating fluid. Hence:

$$C_I = C_M + C_A = c_m \frac{\rho \pi D^2}{4} + \rho A \tag{1.2}$$

where c_m = hydrodynamic coefficient for the section, A = cross-sectional area. Thus the inertia force on the body per unit length can be written:

$$F(t)_{\text{inertia}} = C_I \dot{v} = C_M \dot{v} + C_A \dot{v} \tag{1.3}$$

The value of C_I varies with different section shapes. Even for the same shapes, i.e. circular cylinders, experimental values vary by nearly 50 per cent.

When the movement of the cylinders is to be taken into consideration, equation (1.3) is written as follows:

$$F(t) = C_M (\dot{v} - \ddot{u}) + C_A \dot{v} + C_D (v - \dot{u})|v - \dot{u}| \tag{1.4}$$

The first term is the added or hydrodynamic mass term, and it depends on the motion of the member. The second is the inertial term representing the distortion of the streamlines in the fluid, and is considered to be independent of the acceleration of the structure.

The drag term is the only non-linear term in the above expression. In order to linearise it one can assume that the velocities are distributed as a Gaussian process with zero mean (see Chapter 4). This gives:

$$F(t) = C_M (\dot{v} - \ddot{u}) + C_A \dot{v} + \sqrt{(8/\pi)} \sigma C_D (v - \dot{u}) \tag{1.5}$$

where the standard deviation σ can apply to v or $(v - \dot{u})$. The Gaussian distribution is not justified when working with $(v - \dot{u})$ instead of $\dot{u}$, and in this case one generally uses an iterative procedure to obtain $\sigma_{(v-\dot{u})}$.

The use of the initial $\sigma_{\dot{u}}$ in the calculations, without an iterative improvement, can give rise to errors in the analysis of offshore structures for which the drag effects are significant.

Two sources of uncertainty when applying Morison's equation are the values to be taken for C_I and C_D coefficients. Measures of water particle velocity against wave forces give values of C with a standard deviation of 24 per cent. They also indicate a standard deviation of 22 per cent for the inertia coefficient. Even assuming these comparatively low deviations, one will obtain large variations of response. For instance, taking $C_I \pm 3\sigma_I$ for an inertia-only regime, the response can vary by as much as ± 66 per cent.

A further source of uncertainty is the effect of different wave theories used to calculate the velocities and accelerations used in Morison's equation. Differing theories do not seem to give a large variation in the magnitude of the resulting forces, but they do alter the shape of the force–time curve. Figure 1.4 shows the normalised forces to be expected when using the linear Airy wave theory and the

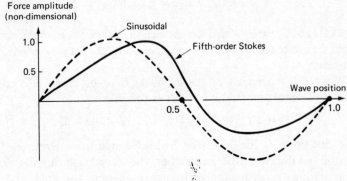

Figure 1.4   Force shape for H = 22.5 m, T = 12 s wave

fifth-order Stokes theory. The shape corresponding to the latter is quite different from a sinusoid, and its asymmetric shape produces secondary peaks in the dynamic response that tend to amplify the effect of smaller waves.

Diffraction is an important effect for large members for which Morison's equation does not apply. It is caused by disturbance of the flow field, owing to the presence of the structure. The structure produces a scattering of the waves, and as these members are in conditions of pure inertia diffraction theory is based on potential flow. The total velocity potential is given as the sum of the incident and diffracted potentials. The two fields are then made to satisfy the boundary conditions. The net effect is a different force distribution from the one that would be obtained if the wave was undisturbed. Figure 1.5 shows the effect in terms of the force spectral density. The effect is important, because most large-diameter structures in 100–150 m water depth have natural frequencies in the range $\omega = 1.5$ (gravity) to 3.0 (steel). Member diameter near the surface may be as large as 10 m for gravity structures and 2 m for steel structures. To give an idea of the importance of the problem, for a 12 m diameter column in 10 m of water, considering or not considering diffraction will show a 100 per cent difference in the calculated deck displacements. The values of the displacements are smaller when considering diffraction, and this is the reason for neglecting it in many cases. Diffraction is, however, of fundamental importance for gravity-structure bases and in other applications such as groups of members, etc. It is generally considered to be a linear problem but is strongly

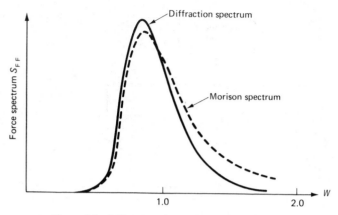

Figure 1.5 Diffraction versus Morison force spectrum

frequency-dependent. Diffraction theory is fully discussed in Chapter 5.

1.4 METHODS OF ANALYSIS

The methods of analysis are probably the more reliable part of the design process, owing to the many recent advances in computational mechanics, but even here a variety of options is open to the analyst. The most important of these possibilities are summarised in Figure 1.6.

The first option is to use a static analysis. One rule of thumb is that dynamic amplification becomes important when the period of the system is larger than 2 seconds. (Note that the period of the structure–water–soil system is generally quite different from the period of the structure only.) It is also important to point out that even structures analysed statically will eventually require some dynamic consideration to analyse the effect of fatigue.

If the inertia forces are comparatively important one uses dynamic analysis, and two possibilities exist. One can work in the frequency domain, in which case transient effects are neglected and one concentrates on steady-state solutions. The method assumes a linear system. The other possibility is to analyse the system in the time domain by some step technique, in which case transient effects may be considered as well as non-linearities.

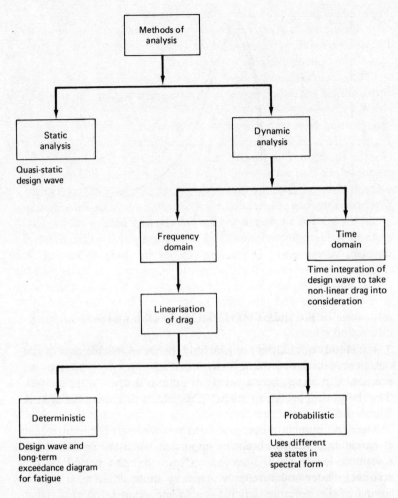

Figure 1.6 Methods of analysis

Owing to computer time savings, the frequency domain analysis is usually preferred, but the time integration method is needed if one wants to study effects such as drag, an earthquake, or non-linear wave theories.

Frequency domain methods can be divided into deterministic and probabilistic methods. Deterministic analysis applies a series of design waves and uses the long-term exceedance diagram for fatigue.

Probabilistic methods have become increasingly popular with analysts during recent years; they can be used to study the behaviour of the structure during the extreme design storm and to study its long-term behaviour for a range of sea states.

Spectral curves of wave height are the information needed to start probabilistic analysis. They are converted into spectral estimates of forces by applying a transfer function, which is simply the square of the resulting wave force per unit wave height. The new spectrum (see Figure 1.7) is then multiplied by the ordinates of the structural transfer function. This transfer function is the square of the dynamic response per unit wave force. The final response curve can then be integrated to produce the variance σ^2, for displacements or stresses. The procedure can be extended to obtain other statistical information about the system, such as long-term behaviour. The drawback is that one analysis is required for each sea state in order to study cumulative damage.

It is nowadays accepted that probabilistic analysis gives good results and is the type of analysis recommended for large structures (see Chapter 9). The main uncertainties in the analysis lie in the estimation of the fatigue life of the structure and the soil–structure interaction effects.

The methods for fatigue analysis may follow either a deterministic or a probabilistic approach. The deterministic approach is based on knowing the height exceedance diagram, usually for one year. Then the relationship between stresses, at the hot spot, and wave height is found, and finally the cumulative stress damage curve is computed by plotting the number of cycles against stress range. The number of cycles can be obtained from the wave exceedance curve. It is generally recommended that a dynamic amplification factor be taken into account. The cumulative stress history curve thus obtained is compared against standard fatigue failure curves. The process is relatively simple and straightforward.

In a probabilistic fatigue analysis one starts by considering the different sea states for a one-year period and the percentage of time during which they act. The number of times that a sea state produces peaks that exceed a particular stress level can be computed from the distribution of stress peaks, as shown in Chapters 2 and 7. The probability curve can then be plotted for each sea state against the stress load. This and the knowledge of the number of cycles per sea state (which is also obtained from statistical considerations) allow the

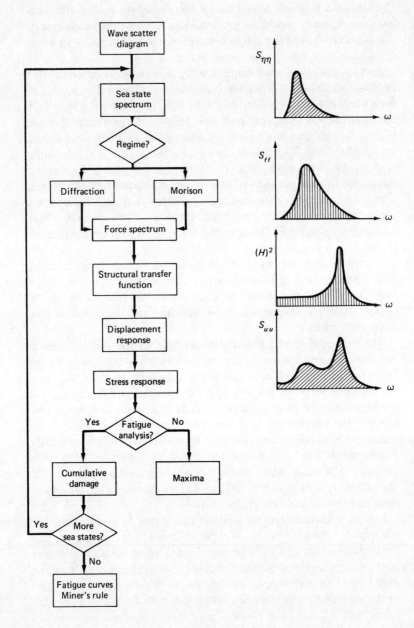

Figure 1.7   Probabilistic analysis

cumulative stress history for a one-year period to be obtained. One can then multiply the results by a stress concentration factor and compare them against standard fatigue failure curves. Hence the fatigue damage for a particular point or 'hot spot' is obtained. The results are influenced by the inaccuracies in the stress concentration factors and the fatigue design curves.

Soil–structure interaction effects are discussed in detail in Chapter 9. The main uncertainties of the analysis are the representation of the material non-linearities and the radiation boundary conditions. The interaction effects produce a different dynamic response for the system, and hence they have to be investigated when dynamic effects are taken into consideration.

2 Fundamentals of Probabilistic Processes

2.1 BASIC CONCEPTS

If the magnitude of a variable associated with a process is known at any moment in time we call it 'deterministic'. The variation of this process may be defined by a known formula or obtained from a graph or experiment.

By contrast, the systems that we are now going to study are subjected to a non-deterministic or 'random' variation of loads. At a given time the variation of the process is not known. What we then do is to define the variation of these processes in terms of some statistical values. This implies that the random records need to be analysed in order to extract from them their basic statistical properties. The system response that these loads produce (displacements, forces, etc.) are also 'random' or statistically defined functions. (Note that the term random does not imply that these forces are arbitrary, but rather that they are statistically defined.)

Figure 2.1 shows a time history of a given process $u(t)$, and demonstrates the difficulty of knowing the precise value of u at a given time t. We can, however, define the probability of u lying within a certain interval. The *probability density* function of a process needs then to be defined as a $p(u)$ function ($\geqslant 0$), such that:

$$\int_{-\infty}^{\infty} p(u)\,\mathrm{d}u = 1 \tag{2.1}$$

14

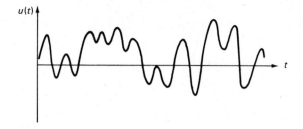

Figure 2.1   Random process

To calculate $p(u)$, consider again a record of a random process of length T and mark the u and $u + du$ lines as shown in Figure 2.2. For T long enough, $p(u)$ is given by:

$$p(u)\,du = \frac{dt_1 + dt_2 + dt_3 + \ldots}{T} = \frac{\sum dt}{T} \qquad (2.2)$$

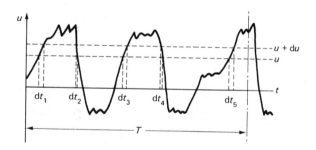

Figure 2.2   Record of length T

In theory the time T is infinite and $p(u)$ tends to a limit. This property will soon be defined in a better way. Once this analysis is carried out we can plot a probability density curve such as the one shown in Figure 2.3.

Most processes in engineering can be approximated by a probability curve called a Gaussian probability curve, and are similar in shape to the one shown in Figure 2.3. Its analytical expression is:

$$p(u) = \frac{1}{\sqrt{(2\pi)}\sigma} \exp\left[-(u - U)^2/2\sigma^2\right] \qquad (2.3)$$

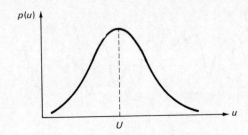

Figure 2.3   Probability density curve

where U and σ are the mean value of u and the standard deviation. The latter is a statistical parameter related to the deviation of the results from the mean value.

Given the probability density function, we can then calculate the probability of a given function u being in a certain interval, e.g. u from u_a to u_b. It is:

$$P(u_a < u < u_b) = \int_{u_a}^{u_b} p(u)\,\mathrm{d}u \tag{2.4}$$

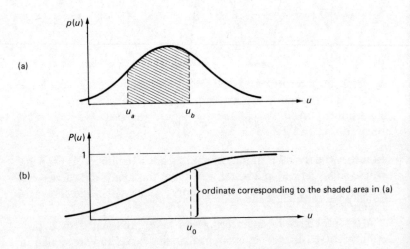

Figure 2.4   Probability density and cumulative probability functions

This represents the area under the probability curve (Figure 2.4) from

u_a to u_b. Note that the value of P will lie between 0 and 1, i.e.

$$P(-\infty < u < \infty) = \int_{-\infty}^{\infty} p(u)\,du = 1 \qquad (2.5)$$

For instance, the probability that the value of u lies between $-\infty$ and u_0 is a cumulative probability function $P(u)$.

$$P(u) = \int_{-\infty}^{u_0} p(u)\,du \qquad (2.6)$$

The $P(u)$ function can also be represented graphically as shown in Figure 2.4, and is called the cumulative probability function. Note that the relationship between P and p can also be expressed as:

$$\frac{dP(u)}{du} = p(u) \qquad (2.7)$$

The *mean value* of a process u is denoted by $\langle u \rangle$, where $\langle\ \rangle$ represents the statistical expectation of u and is defined by:

$$\langle u \rangle = \lim_{T \to \infty} \frac{1}{T} \int_0^T u(t)\,dt \qquad (2.8)$$

We can also write this:

$$\langle u \rangle = \lim_{T \to \infty} \int_0^T u(t)\,\frac{dt}{T} \qquad (2.9)$$

dt/T can be interpreted as the fraction of time during which the process is between the two values u and $u + du$. This is also given by the product $p(u)\,du$. We can then pass from an integral in time to an integral in u, and define the mean as:

$$\langle u \rangle = \int_{-\infty}^{\infty} u\,p(u)\,du \qquad (2.10)$$

Equally we can define the *mean square value* of u as:

$$\langle u^2 \rangle = \lim_{T \to \infty} \frac{1}{T} \int_0^T u^2\,dt \qquad (2.11)$$

or as the second moment of the probability function, i.e.

$$\langle u^2 \rangle = \int_{-\infty}^{\infty} u^2 p(u)\,du \qquad (2.12)$$

The *variance* is defined as:

$$\sigma^2 = \langle (u - \langle u \rangle)^2 \rangle \tag{2.13}$$

This can be simplified by expanding as:

$$\sigma^2 = \langle u^2 - 2u \langle u \rangle + \langle u \rangle^2 \rangle$$
$$= \langle u^2 \rangle - 2 \langle u \rangle \langle u \rangle + \langle u \rangle^2 \tag{2.14}$$
$$\sigma^2 = \langle u^2 \rangle - \langle u \rangle^2$$

i.e.   Variance = Mean square − square of mean (2.15)

The *standard deviation*, σ, is defined as the positive square root of σ^2, the variance.

As an application, consider the mean value of a Gaussian process:

$$\langle u \rangle = \int_{-\infty}^{\infty} u p(u) \, du = \int_{-\infty}^{\infty} u \frac{\exp[-(u - U)^2/2\sigma^2]}{\sqrt{(2\pi)}\sigma} \, du \tag{2.16}$$

This integral can be evaluated by changing the variable of integration to $\xi = u - U$. Hence:

$$\langle u \rangle = \frac{1}{\sqrt{(2\pi)}\sigma} \int_{-\infty}^{\infty} (\xi + U) \exp(-\xi^2/2\sigma^2) \, d\xi \tag{2.17}$$

From tables*:

$$\langle u \rangle = U \tag{2.18}$$

We can also find:

$$\langle u^2 \rangle = \frac{1}{\sqrt{(2\pi)}\sigma} \int_{-\infty}^{\infty} u^2 \exp[-(u - U)^2/2\sigma^2] \, du \tag{2.19}$$

With the same change of variables ($\xi = u - U$) we obtain:

$$\langle u^2 \rangle = \frac{1}{\sqrt{(2\pi)}\sigma} \int_{-\infty}^{\infty} (\xi + U)^2 \exp(-\xi^2/2\sigma^2) \, d\xi \tag{2.20}$$

which gives:

$$\langle u^2 \rangle = \sigma^2 + U^2 = \sigma^2 + \langle u \rangle^2 \tag{2.21}$$

---

* $\int_0^\infty \exp(-\xi^2/2\sigma^2) \, d\xi = \sqrt{(\pi/2)}\sigma, \quad \int_0^\infty \xi \exp(-\xi^2/2\sigma^2) \, d\xi = \sigma^2,$

$\int_0^\infty \xi^2 \exp(-\xi^2/2\sigma^2) \, d\xi = \sqrt{(\pi/2)}\sigma^3$

The values define the mean, $\langle u \rangle$, and standard deviation, σ, of a Gaussian process.

In many practical applications a physical problem can be studied by breaking the physical record down into two parts:

- a stationary part, invariant in time and defined by the mean value;
- a dynamic part, whose fluctuations around the mean are defined by σ.

One can then carry out two different analyses, i.e. static and dynamic analysis. The static case is studied using the mean value. For the dynamic case, the mean is taken to be zero and the mean square value is:

$$\sigma^2 = \langle u^2 \rangle \tag{2.22}$$

Example 2.1

We shall now calculate the probability density function, mean and standard deviation for the deterministic variable given by:

$$u(t) = a \sin \omega t \quad \text{(Figure 2.5)} \tag{a}$$

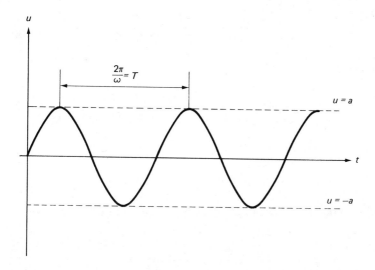

Figure 2.5 $u = a \sin \omega t$

From (a) we obtain:

$$du = a\omega \cos \omega t \times dt \tag{b}$$

So:

$$dt = \frac{du}{a\omega \cos \omega t} \tag{c}$$

or:

$$dt = \frac{du}{a\omega\sqrt{(1 - u^2/a^2)}} \tag{d}$$

The proportion of time per cycle that $u(t)$ spends in the range $\hat{u}$ to $\hat{u} + du$ is then:

$$\frac{2dt}{T} = \frac{2du}{\omega T\sqrt{(a^2 - \hat{u}^2)}} \tag{e}$$

where T is the period of one cycle. Using:

$$T = \frac{2\pi}{\omega} \tag{f}$$

(e) becomes:

$$\frac{2dt}{T} = \frac{du}{\pi\sqrt{(a^2 - \hat{u}^2)}} \tag{g}$$

Hence:

$$\text{Prob}(\hat{u} \leqslant u \leqslant \hat{u} + du) = \frac{2dt}{T} = \frac{du}{\pi\sqrt{(a^2 - \hat{u}^2)}} \tag{h}$$

for $-a \leqslant \hat{u} \leqslant a$. But:

$$\text{Prob}(\hat{u} \leqslant u \leqslant \hat{u} + du) = p(\hat{u})\,du \tag{i}$$

where $p(\hat{u})$ is the probability density function evaluated at $u = \hat{u}$.

$$\therefore \qquad p(u) = \frac{1}{\pi\sqrt{(a^2 - u^2)}} \tag{j}$$

This function is illustrated in Figure 2.6. $P(u)$ is zero for $|u| > a$.

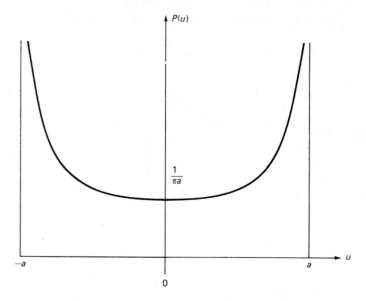

Figure 2.6   Probability density function for $u = a \sin \omega t$

The probability that u lies between u_a and u_b is:

$$\text{Prob}(u_a \leqslant u \leqslant u_b) = \int_{u_a}^{u_b} p(u)\,du \qquad \text{(k)}$$

$$= \int_{u_a}^{u_b} \frac{du}{\pi\sqrt{(a^2 - u^2)}} \qquad \text{(l)}$$

$$= \frac{1}{\pi}\left[\sin^{-1}\left(\frac{u_b}{a}\right) - \sin^{-1}\left(\frac{u_a}{a}\right)\right] \qquad \text{(m)}$$

We see from the symmetry of $\sin \omega t$ about the t axis that the expected value of u, $\langle u \rangle$ will be zero. Indeed:

$$\langle u \rangle = \lim_{T \to \infty} \int_0^T \sin \omega t\, \frac{dt}{T} = \int_{-\infty}^{\infty} up(u)\,du = 0 \qquad \text{(n)}$$

as the integrand is an odd function of u.

The mean square value (or variance of u) is given by:

$$\langle u^2 \rangle = \lim_{T \to \infty} \int_0^T \sin^2 \omega t \, \frac{dt}{T} = \int_{-\infty}^{\infty} u^2 p(u) \, du \qquad (o)$$

$$= \int_{-u}^{u} \frac{u^2 \, du}{\pi \sqrt{(a^2 - u^2)}} \qquad (p)$$

as $p(u)$ is zero for $|u| > a$. Hence:

$$\sigma_u^2 = \langle u^2 \rangle = \frac{a^2}{2} \qquad (q)$$

by the substitution $u = a \sin \theta$, and the standard deviation is:

$$\sigma_u = \frac{a}{\sqrt{2}}$$

Stationary processes

Let us now define two important properties of the records under consideration. Take the case of several records taken for the same variable (for instance, the records corresponding to the movements of the wing tip of an airplane during flight obtained for a number of different flights; Figure 2.7). Note that if the number of records is large enough one can calculate ensemble averages across the ensemble, as well as time averages.

A process is called *stationary* if its probability distribution does not depend on time. This implies that the mean, mean square, variance and standard deviation of the process are independent of time. This is a condition that can only be approximately satisfied in engineering processes for which there is always a beginning and an end, but in most cases it is reasonable to assume stationarity. For the case shown in Figure 2.7 we would have that the probability distribution function of each of the records u_i does not depend on the actual instant chosen as the origin of time.

If in addition the probability distribution is the same for each of the records, the stationary process is said to be *ergodic*. In what follows we will consider only *stationary ergodic* processes, which have the advantage that all properties of the process can be obtained from a single record.

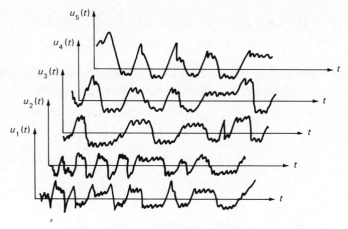

Figure 2.7   Several records

The classic example of a stationary process is wing-tip deflection during flight between two distant cities. We can take the records corresponding to the airplane flying in level flight (i.e. neglecting take-off and landing) during days for which the atmospheric conditions are similar. If one analyses one record starting at different times and finds always the same probability distribution, the flight record is said to be stationary. Assuming that this is true for all the other records, one can now compare their probability distributions; if they are the same, the process (flights between the two cities) is said to be *ergodic* as well as *stationary* and is completely defined by the properties of one record. Note that, although all these properties apply for processes extending from $t = -\infty$ to $t = +\infty$, we can assume that they are still valid for sufficiently long records.

2.2 SECOND-ORDER PROBABILITY FUNCTIONS

A second-order probability function $p(u, v)$ depends on two random variables u and v. The probability that u lies in the range u to $u + du$, and v in the range v to $v + dv$, is given by $p(u, v)\,du\,dv$.

If we assume we have two bands u_1 to u_2 and v_1 to v_2, the probability of u and v lying within these two regions is given by the following integral:

$$\text{Prob}(u_1 \leqslant u \leqslant u_2 \text{ and } v_1 \leqslant v \leqslant v_2) = \int_{u_1}^{u_2} \int_{v_1}^{v_2} p(u, v) \, \mathrm{d}u \, \mathrm{d}v$$

$$(2.23)$$

If the limits tend to $-\infty$ and ∞, the probability becomes 1 (i.e. the random variables u and v will be within the integral limits).

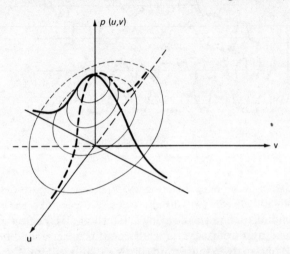

Figure 2.8   Second-order probability density function

$p(u, v)$ can be represented as a two-dimensional surface containing a volume of 1 (Figure 2.8). The first-order probability function can be obtained from the second-order function by integrating $p(u, v)$ within a band $\mathrm{d}v$, and letting the limits tend to $-\infty$ and ∞, i.e.

$$p(u) = \int_{-\infty}^{\infty} p(u, v) \, \mathrm{d}v \qquad (2.24)$$

The expectation or average value of a quantity f, a function of two variables u and v, is:

$$\langle f(u, v) \rangle = \int_{-\infty}^{\infty} \int_{-\infty}^{\infty} f(u, v) \, p(u, v) \, \mathrm{d}u \, \mathrm{d}v \qquad (2.25)$$

If u and v are *independent*, their joint probability density function can be written as a function of the probability of u only and probability of v only, i.e.

$$p(u, v) = p(u) \cdot p(v) \qquad (2.26)$$

The two-dimensional Gaussian distribution can generally be written as:

$$p(u, v) = \frac{1}{2\pi\sigma_u\sigma_v\sqrt{(1-\rho^2)}} \times$$

$$\times \exp\left\{-\frac{1}{2(1-\rho^2)}\left[\frac{(u-U)^2}{\sigma_u^2} + \frac{(v-V)^2}{\sigma_v^2} - \frac{2\rho(u-U)(v-V)}{\sigma_u\sigma_v}\right]\right\}$$

$$(2.27)$$

where U and V are the mean values of u and v; ρ is a correlation coefficient defined as:

$$\rho = \frac{\langle(u-U)(v-V)\rangle}{\sigma_u\sigma_v} \qquad (2.28)$$

For the case of u and v being independent we have $\rho = 0$, and the probability density function becomes:

$$p(u, v) = \frac{1}{\sqrt{(2\pi)}\sigma_u}\exp\left[-\frac{(u-U)^2}{2\sigma_u^2}\right] \cdot \frac{1}{\sqrt{(2\pi)}\sigma_v}\exp\left[-\frac{(v-V)^2}{2\sigma_v^2}\right]$$

$$= p(u) \cdot p(v) \qquad (2.29)$$

It is interesting to note that the case $\rho = \pm 1$ represents perfect correlation, and we have:

$$p(u, v) = \infty \qquad \text{for } u = v \qquad (2.30)$$

2.3 n-DIMENSIONAL PROBABILITY DENSITY FUNCTIONS

A more general expression than the two-dimensional Gaussian probability distribution is the one for n dimensions. Before writing the general expression, consider the case of only two variables and define for them a correlation matrix:

$$\mathbf{\mu} = \begin{bmatrix} \mu_{11} & \mu_{12} \\ \mu_{21} & \mu_{22} \end{bmatrix} = \begin{bmatrix} \langle u-U, u-U\rangle & \langle u-U, v-V\rangle \\ \langle v-V, u-U\rangle & \langle v-V, v-V\rangle \end{bmatrix}$$

$$= \begin{bmatrix} \sigma_u^2 & \langle u-U, v-V\rangle \\ \langle v-V, u-U\rangle & \sigma_v^2 \end{bmatrix} \qquad (2.31)$$

The determinant of this matrix is:

$$|\mu| = \sigma_u^2\sigma_v^2 - \langle u-U, v-V\rangle^2 \qquad (2.32)$$

and the inverse of $\boldsymbol{\mu}$ can be written as:

$$\boldsymbol{\mu}^{-1} = \begin{bmatrix} \sigma_v^2 & -\langle u-U, v-V \rangle \\ -\langle u-U, v-V \rangle & \sigma_u^2 \end{bmatrix} \frac{1}{\sigma_u^2 \sigma_v^2 - \langle u-U, v-V \rangle^2}$$

(2.33)

We can also define a vector $(\mathbf{x} - \mathbf{X})$ such that:

$$(\mathbf{x} - \mathbf{X}) = \begin{Bmatrix} u - U \\ v - V \end{Bmatrix}$$

(2.34)

It is now easy to verify that the equation for $p(u, v)$ can be written as:

$$p(u, v) = \frac{1}{2\pi |\boldsymbol{\mu}|^{\frac{1}{2}}} \exp\left[-\tfrac{1}{2}(\mathbf{x} - \mathbf{X})^T \boldsymbol{\mu}^{-1}(\mathbf{x} - \mathbf{X}) \right]$$

(2.35)

We can generalise this result for n variables using x_i instead of u, v. This gives the following n-dimensional expression:

$$p(x_1 x_2 \ldots x_n) = \frac{1}{(2\pi)^{n/2} |\boldsymbol{\mu}|^{\frac{1}{2}}} \exp\left[-\tfrac{1}{2}(\mathbf{x} - \mathbf{X})^T \boldsymbol{\mu}^{-1}(\mathbf{x} - \mathbf{X}) \right]$$

(2.36)

The components of the correlation matrix are:

$$\mu_{ij} = \langle (x_i - X_i)(x_j - X_j) \rangle$$

(2.37)

2.4 AUTOCORRELATION AND SPECTRAL DENSITY FUNCTIONS

Autocorrelation

The autocorrelation function for a process $u(t)$ is the average value of the product $u(t)u(t + \tau)$, where τ is a time lag and is defined as:

$$R(\tau) = \lim_{T \to \infty} \frac{1}{T} \int_0^T u(t)u(t + \tau)\,\mathrm{d}t$$

(2.38)

This function gives some information about the value of a signal at the instant $t + \tau$, when the value at t is known. Note that for $\tau = 0$ we have the mean square value:

$$R(0) = \langle u^2 \rangle$$

(2.39)

As τ gets longer the autocorrelation will decrease. If $R(\tau)$ decreases rapidly we have a disconnected process.

Example 2.2

We shall now calculate the autocorrelations and cross-correlations of the two variables u and v given by:

$$u(t) = a \sin \omega t \tag{a}$$

$$v(t) = b \sin (\omega t + \alpha) \tag{b}$$

u and v are sine waves of amplitude a and b, with circular frequency ω; u and v are out of phase with each other by a fixed angle α.

The autocorrelation function for u is given by:

$$R_{uu}(\tau) = \lim_{T \to \infty} \frac{1}{T} \int_0^T u(t)\,u(t+\tau)\,dt \tag{c}$$

$$= \lim_{T \to \infty} \frac{1}{T} \int_0^T a \sin \omega t\, a \sin\left[\omega(t+\tau)\right]dt \tag{d}$$

$$= \lim_{T \to \infty} \frac{1}{T} a^2 \int_0^T \{\sin \omega t[\sin \omega t \cos \omega t] +$$

$$+ \cos \omega t[\sin^2 \omega t]\}\,dt \tag{e}$$

$$= \lim_{T \to \infty} a^2 \left[\sin \omega t \left(\frac{1}{2T} \frac{\cos 2\omega T}{2\omega T} \right) + \right.$$

$$\left. + \cos \omega t \left(1 - \frac{1}{T} \frac{\sin 2\omega T}{2\omega T} \right) \right] \tag{f}$$

On taking the limit, the terms involving T go to zero as $|\cos 2\omega T| < 1$ always. Hence we obtain:

$$R_{uu}(\tau) = \frac{a^2}{2} \cos \omega \tau \tag{g}$$

The autocorrelation function for v is calculated similarly and is given by:

$$R_{vv}(\tau) = \lim_{T \to \infty} \frac{1}{T} \int_0^T v(t)\,v(t+\tau)\,dt = \frac{b^2}{2} \cos (\omega \tau) \tag{h}$$

The autocorrelation functions given by (g) and (h) express the fact that the functions are well correlated for time separations of $2\pi n/\omega = nT$, i.e. the original variables are periodic with period T.

The cross-correlation function between u and v is now:

$$R_{uv}(\tau) = \lim_{T \to \infty} \frac{1}{T} \int_0^T ab \sin \omega t \sin [\omega(t+\tau) + \alpha] \, dt \tag{i}$$

$$= \lim_{T \to \infty} \frac{ab}{T} \int_0^T \{\cos(\omega\tau + \alpha) [\sin^2 \omega t] + \sin(\omega\tau + \alpha) \times$$

$$\times [\sin \omega t \cos \omega t]\} \, dt \tag{j}$$

$$= \frac{ab}{2} \cos(\omega\tau + \alpha) \tag{k}$$

This function is represented in Figure 2.9.

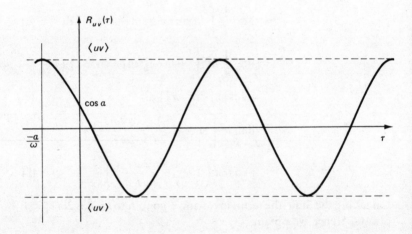

Figure 2.9   Cross-correlation function for two sine waves with $v(t)$ leading $u(t)$ by angle α

The cross-correlation function between v and u is given by:

$$R_{vu}(\tau) = \frac{ab}{2} \cos(\omega\tau - \alpha) \tag{l}$$

The angle α is now treated as a phase lag. We see that:

$$R_{uu}(0) = \frac{a^2}{2} = \langle u^2 \rangle \tag{m}$$

$$R_{vv}(0) = \frac{b^2}{2} = \langle v^2 \rangle \tag{n}$$

$$R_{uv}(0) = \frac{ab}{2} \cos \alpha = R_{vu}(0) = \langle uv \rangle = \langle vu \rangle \tag{o}$$

in accordance with equation (2.39).

Spectral density

Before defining the spectral density, let us consider a function $u(t)$ which can be represented by a Fourier series:

$$u(t) = a_0 + \sum_{n=1}^{\infty} a_n \cos n\omega t + \sum_{n=1}^{\infty} b_n \sin n\omega t \tag{2.40}$$

where T is the period of the function and $\omega = 2\pi/T$ is the lowest frequency corresponding to the interval T.

The constants are evaluated by the following integrals:

$$a_0 = \frac{1}{T} \int_{-T/2}^{T/2} u(t)\, dt$$

$$a_n = \frac{2}{T} \int_{-T/2}^{T/2} u(t) \cos n\omega t\, dt \qquad n \geqslant 1 \tag{2.41}$$

$$b_n = \frac{2}{T} \int_{-T/2}^{T/2} u(t) \sin n\omega t\, dt$$

If the mean value of $u(t)$ is zero, then $a_0 = 0$. The a_n and b_n coefficients can be represented as a series of ordinates with a given ω spacing (Figure 2.10). When the period T increases, the frequency ω will become smaller and the ordinates in Figure 2.10 become very near one another; for $T \to \infty$ they will join together. For this case we will see that the u function has to be described by a Fourier integral instead of by a discrete Fourier series.

A more general way of representing a Fourier series is by using complex notation. This can be done by substituting

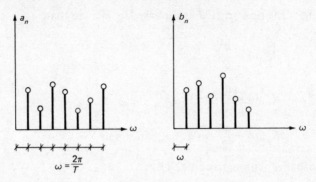

Figure 2.10   Frequency-domain representation of the Fourier series function

$\cos n\omega t = \frac{1}{2}[\exp(in\omega t) + \exp(-in\omega t)]$ and $\sin n\omega t = [\exp(in\omega t)$
$- \exp(-in\omega t)]/2i$, giving:

$$u(t) = a_0 + \sum_{n=1}^{\infty} [c_n \exp(in\omega t) + d_n \exp(-in\omega t)] \qquad n = 1, 2 \ldots$$

$$(2.42)$$

where $c_n = (a_n - ib_n)/2$, $d_n = (a_n + ib_n)/2$. The same coefficients can also be written as:

$$c_n = \frac{1}{T} \int_{-T/2}^{T/2} u(t) \exp(-in\omega t)\, dt$$
$$\qquad n = 1, 2 \ldots (2.43)$$
$$d_n = \frac{1}{T} \int_{-T/2}^{T/2} u(t) \exp(in\omega t)\, dt$$

We can now introduce the notation $d_n = c_{-n}$, which means that all the coefficients are defined by c_n, i.e.

$$u(t) = \sum_{n=-\infty}^{\infty} c_n \exp(in\omega t) \qquad n = 0, \pm 1, \pm 2 \ldots \qquad (2.44)$$

This is the complex form of Fourier series, with c_n the complex Fourier coefficients of $u(t)$.

Substituting c_n into equation (2.44), we can write:

$$u(t) = \omega \sum_{n=-\infty}^{\infty} \left[\frac{1}{2\pi} \int_{-T/2}^{T/2} u(t) \exp(-in\omega t)\, dt \right] \exp(in\omega t)$$

$$= \omega \sum_{n=-\infty}^{\infty} F(n\omega)\exp(in\omega t) \qquad (2.45)$$

where $1/T$ has been replaced by $\omega/2\pi$.

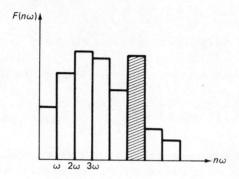

Figure 2.11   *Discrete representation of F (nω) function*

The $F(n\omega)$ can now be interpreted as the ordinates of a function for different values of n ($n, 2n, 3n, 4n$, etc.), as shown in Figure 2.11. Note that each of the areas in the figure has a base ω and an ordinate F. When ω becomes smaller we can replace it by $\delta\omega$ and the values on the abscissa are $\delta\omega, 2\delta\omega, 3\delta\omega \ldots$ etc. If now $\delta\omega \to 0$ (and $T \to \infty$), we have a continuous variation of ω and we can replace the summation by an integral to obtain:

$$u(t) = \int_{-\infty}^{\infty} \frac{1}{2\pi} \left[\int_{-\infty}^{\infty} u(t)\exp(-i\omega t)\,dt \right] \exp(i\omega t)\,d\omega \qquad (2.46)$$

Note that the integral inside the parenthesis is a function of t and can be called $\overline{U}$ or the Fourier transform of u.

$$\overline{U}(\omega) = \frac{1}{2\pi} \int_{-\infty}^{\infty} u(t)\exp(-i\omega t)\,dt \qquad (2.47)$$

Hence we can now write u as:

$$u(t) = \int_{-\infty}^{\infty} \overline{U}(\omega)\exp(i\omega t)\,d\omega \qquad (2.48)$$

$u(t)$ and $\overline{U}(\omega)$ are said to form a Fourier transform pair. (The position of the 2π in $\overline{U}$ is quite arbitrary and different authors prefer to put it in one or the other of the integrals. The notation used here is the one most widely adopted.)

Consider now the mean square value of the process $u(t)$, i.e.

$$\langle u^2 \rangle = \lim_{T \to \infty} \frac{1}{T} \int_{-T/2}^{T/2} u(t)\,u(t)\,dt \tag{2.49}$$

By the previous definitions of Fourier transforms:

$$\langle u^2 \rangle = \lim_{T \to \infty} \frac{1}{T} \int_{-T/2}^{T/2} u(t) \left[\int_{-\infty}^{\infty} \overline{U}(\omega)\,\exp(i\omega t)\,d\omega \right] dt$$

$$= \int_{-\infty}^{\infty} \overline{U}(\omega) \left[\lim_{T \to \infty} \frac{1}{T} \int_{-T/2}^{T/2} u(t)\exp(i\omega t)\,dt \right] d\omega \tag{2.50}$$

The integral in parenthesis is the Fourier transform of $u(t)$ but with $+i$ instead of $-i$; i.e. it is the complex conjugate of $\overline{U}$, which we shall call $\hat{U}$ in what follows. Notice that, when $T \to \infty$, $\hat{U}$ will also tend to infinity. We can now write:

$$\langle u^2 \rangle = \lim_{T \to \infty} \frac{1}{T} \int_{-\infty}^{\infty} \overline{U}\hat{U}\,d\omega$$

$$= \lim_{T \to \infty} \frac{1}{T} \int_{-\infty}^{\infty} |\overline{U}|^2\,d\omega \tag{2.51}$$

$$= \int_{-\infty}^{\infty} \left(\lim_{T \to \infty} \frac{1}{T} |\overline{U}|^2 \right) d\omega \tag{2.52}$$

The term between brackets is called the spectral density function for the u process.

$$S_{uu}(\omega) = \lim_{T \to \infty} \frac{1}{T} |\overline{U}|^2 \tag{2.53}$$

Notice that $S_{uu}(\omega)$ can be interpreted as the contribution to $\langle u^2 \rangle$ associated with the frequency ω, and is analogous to having a continuous variation of the coefficients of a Fourier series.

2.5 RELATIONSHIP BETWEEN AUTOCORRELATION AND SPECTRAL DENSITY

Consider again the definition of autocorrelation for the process u:

$$R_{uu}(\tau) = \lim_{T \to \infty} \frac{1}{T} \int_0^T u(t)\,u(t+\tau)\,dt \qquad (2.54)$$

Notice now that the Fourier transform pair for a function such as $u(t+\tau)$ will be:

$$\bar{U} = \frac{1}{2\pi} \int_{-\infty}^{\infty} u(t+\tau)\exp(-i\omega t)\,dt\,\exp(-i\omega\tau)$$

$$u(t+\tau) = \int_{-\infty}^{\infty} \bar{U}\exp[i\omega(t+\tau)]\,d\omega \qquad (2.55)$$

These relationships are valid because the starting time can be taken as t or $t+\tau$ in a stationary process without any changes.

The autocorrelation can now be written as:

$$R_{uu}(\tau) = \lim_{T \to \infty} \frac{1}{T} \int_{-T/2}^{T/2} u(t) \int_{-\infty}^{\infty} \bar{U}(\omega)\ \exp[i\omega(t+\tau)]\,d\omega dt$$

$$= \int_{-\infty}^{\infty} \lim_{T \to \infty} \frac{1}{T}\left[\bar{U}(\omega) \int_{-T/2}^{T/2} u(t)\exp(i\omega t)\,dt \right]\exp(i\omega\tau)\,d\omega \qquad (2.56)$$

$$= \int_{-\infty}^{\infty} \lim_{T \to \infty} \frac{1}{T}\ \bar{U}(\omega)\,\hat{U}(\omega)\exp(i\omega\tau)\,d\omega$$

$$= \int_{-\infty}^{\infty} S_{uu}(\omega)\exp(i\omega\tau)\,d\omega$$

This implies that the autocorrelation $R_{uu}(\tau)$ is the Fourier transform of the spectral density of the process u. Similarly the spectral density can be expressed as:

$$S_{uu}(\omega) = \frac{1}{2\pi} \int_{-\infty}^{\infty} R_{uu}(\tau)\exp(-i\omega\tau)\,d\tau \qquad (2.57)$$

These last two formulae are called the Weiner–Khichin relationships.

Note that the autocorrelation and the spectral density are even functions of τ and ω respectively.

$$R(\tau) = R(-\tau)$$

$$S(\omega) = S(-\omega) \qquad (2.58)$$

Hence we can write:

$$R_{uu}(\tau) = 2 \int_0^\infty S_{uu}(\omega) \exp(+i\omega\tau)\,d\omega$$

$$S_{uu}(\omega) = \frac{1}{\pi} \int_0^\infty R_{uu}(\tau) \exp(-i\omega\tau)\,d\tau$$

$$(2.59)$$

It is most important to note that for $\tau = 0$ we have the following result:

$$R_{uu}(0) = 2 \int_0^\infty S_{uu}(\omega)\,d\omega = \langle u^2 \rangle \qquad (2.60)$$

that is, the mean square value of a stationary process is the area under the $S_{uu}(\omega)$ curve (Figure 2.12).

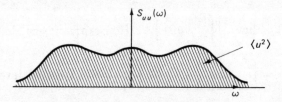

Figure 2.12   Spectral density

Notice also that the integrals from $-\infty$ to 0 and the one from 0 to ∞ are equal. Because of this one never works in practice with negative frequencies, and the spectral density is defined as twice the spectral density of the last formula. For the moment we will continue working with negative frequencies, as they are mathematically more convenient.

2.6 NARROW-BAND AND BROAD-BAND PROCESSES

A process for which the spectral density function is mainly within a narrow band of frequencies is called a *narrow-band process*. If the opposite occurs, i.e. the spectral density is distributed within a broad band of frequencies, the process is called *broad banded*. See Figure 2.13.

The limit of this type of process is the so-called white-noise

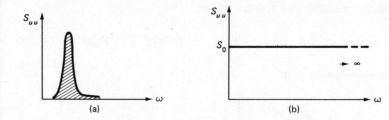

Figure 2.13 Spectra: (a) narrow-band, (b) broad-band

spectrum, which has a constant value of S from $\omega = -\infty$ to $\omega = +\infty$. Notice that the mean square value of a white-noise spectrum is infinite, which makes this choice of spectrum highly suspect. The idea of a white-noise spectrum, however, is useful in practical applications. The autocorrelation function for a white-noise spectrum can be obtained by obtaining the transformation:

$$R_{uu}(\tau) = \int_{-\infty}^{\infty} S_0 \exp(+i\omega\tau) \, \mathrm{d}\omega \tag{2.61}$$

which gives a δ function at $\tau = 0$:

$$R_{uu}(\tau) = 2\pi S_0 \, \delta(\tau) \tag{2.62}$$

Once the spectral density of a process u is known, we can calculate the spectral density of the velocity, $\dot{u} = \mathrm{d}u/\mathrm{d}t$, and acceleration, $\ddot{u} = \mathrm{d}^2u/\mathrm{d}t^2$, processes. Consider:

$$R_{uu}(\tau) = \int_{-\infty}^{\infty} u(t) u(t+\tau) \, \mathrm{d}t \tag{2.63}$$

$$\frac{\mathrm{d}R_{uu}(\tau)}{\mathrm{d}\tau} = \int_{-\infty}^{\infty} u(t) \frac{\mathrm{d}u}{\mathrm{d}\tau}(t+\tau) \, \mathrm{d}t$$

$$= \int_{-\infty}^{\infty} u(t) \frac{\mathrm{d}u(t+\tau)}{\mathrm{d}(t+\tau)} \overbrace{\frac{\mathrm{d}(t+\tau)}{\mathrm{d}\tau}}^{=1} \, \mathrm{d}t$$

$$= \int_{-\infty}^{\infty} u(t) \dot{u}(t+\tau) \, \mathrm{d}t \tag{2.64}$$

Note that for a stationary process one can also write:

$$\frac{\mathrm{d}R_{uu}(\tau)}{\mathrm{d}\tau} = \int_{-\infty}^{\infty} u(t-\tau) \dot{u}(t) \, \mathrm{d}t \tag{2.65}$$

Differentiating with respect to time:

$$\frac{d^2}{d\tau^2} R_{uu}(\tau) = - \int_{-\infty}^{\infty} \dot{u}(t-\tau)\dot{u}(t)\,dt = -R_{\dot{u}\dot{u}}(\tau) \qquad (2.66)$$

We also know that:

$$R_{uu}(\tau) = \int_{-\infty}^{\infty} S_{uu}(\omega)\exp{(i\omega\tau)}\,d\omega \qquad (2.67)$$

$$\therefore \qquad \frac{d^2}{d\tau^2} R_{uu}(\tau) = - \int_{-\infty}^{\infty} \omega^2 S_{uu}(\omega)\exp{(i\omega\tau)}\,d\omega \qquad (2.68)$$

Hence:

$$R_{\dot{u}\dot{u}}(\tau) = \int_{-\infty}^{\infty} \omega^2 S_{uu}(\omega)\exp{(i\omega\tau)}\,d\omega$$

$$= \int_{-\infty}^{\infty} S_{\dot{u}\dot{u}}(\omega)\exp{(i\omega\tau)}\,d\omega \qquad (2.69)$$

Therefore:

$$S_{\dot{u}\dot{u}}(\omega) = \omega^2 S_{uu}(\omega) \qquad (2.70)$$

Similarly we can find that:

$$R_{\ddot{u}\ddot{u}}(\tau) = \int_{-\infty}^{\infty} \omega^4 S_{uu}(\omega)\exp{(i\omega\tau)}\,d\omega \qquad (2.71)$$

and that:

$$S_{\ddot{u}\ddot{u}}(\omega) = \omega^4 S_{uu}(\omega) \qquad (2.72)$$

It can be shown that the cross-correlation between u and $\dot{u}$ is always zero. Consider:

$$R_{u\dot{u}}(\tau) = \int_{-\infty}^{\infty} S_{u\dot{u}}(\omega)\exp{(i\omega\tau)}\,d\omega \qquad (2.73)$$

It is also possible to write this as:

$$\frac{d}{d\tau}[R_{uu}(\tau)] = i \int_{-\infty}^{\infty} \omega S_{uu}(\omega)\exp{(i\omega\tau)}\,d\omega \qquad (2.74)$$

The integrand is an *odd* function of ω; hence integration from $-\infty$ to 0 and 0 to ∞ produces the same term with a different sign, i.e. it cancels.

The variances of u, $\dot{u}$ and $\ddot{u}$, for a zero mean process, are:

$$\sigma_{uu}^2 = \int_{-\infty}^{\infty} S_{uu}(\omega)\,d\omega$$

$$\sigma_{\dot{u}\dot{u}}^2 = \int_{-\infty}^{\infty} \omega^2 S_{uu}(\omega)\,d\omega \qquad (2.75)$$

$$\sigma_{\ddot{u}\ddot{u}}^2 = \int_{-\infty}^{\infty} \omega^4 S_{uu}(\omega)\,d\omega$$

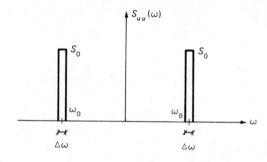

Figure 2.14 Narrow-band process

For a narrow-band process like the one shown in Figure 2.14, these values are:

$$\sigma_{uu}^2 \simeq 2S_0\,\Delta\omega$$

$$\sigma_{\dot{u}\dot{u}}^2 \simeq 2S_0\,\omega_0^2\,\Delta\omega \qquad (2.76)$$

$$\sigma_{\ddot{u}\ddot{u}}^2 \simeq 2S_0\,\omega_0^4\,\Delta\omega$$

Hence the characteristic frequency ω_0 for this case will be:

$$\omega_0^2 = \frac{\sigma_{\dot{u}\dot{u}}^2}{\sigma_{uu}^2} \qquad (2.77)$$

If the *moments* of the spectrum are defined as:

$$M_n = \int_{-\infty}^{\infty} S_{uu}(\omega)\omega^n\,d\omega \qquad (2.78)$$

the characteristic frequency can be written as:

$$\omega_0^2 = \frac{M_2}{M_0} \qquad (2.79)$$

This result is only valid for narrow-band processes.

Many engineering processes are narrow-band, but in general they vary between infinitely narrow and infinitely broad (i.e. white noise) types. A parameter that defines the spectral width is ε:

$$\varepsilon = \left(1 - \frac{M_2^2}{M_0 M_4}\right)^{\frac{1}{2}} \tag{2.80}$$

where $\varepsilon = 0$ for an infinitely narrow spectrum, 1 for an infinitely broad spectrum. A way of calculating this parameter will be presented later.

Example 2.3

We shall now examine the autocorrelation function and expectation of a narrow-band process. We shall start with the (two sided) spectral density function illustrated in Figure 2.15. This function is zero except when $\omega_1 < \omega < \omega_2$ or $-\omega_2 < \omega < \omega_1$ and $a^2/4$ in these ranges.

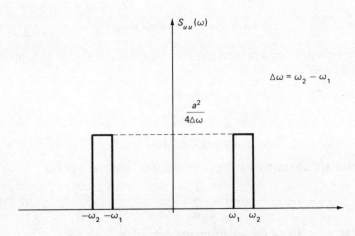

Figure 2.15   Spectral density of narrow-band process

From equations (2.59) we have:

$$R_{uu}(\tau) = \int_{-\infty}^{\infty} S_{uu}(\omega) \exp(i\omega\tau)\, d\omega = \int_{-\infty}^{\infty} S_{uu}(\omega) \cos \omega\tau\, d\omega \tag{a}$$

$$= \frac{a^2}{4\Delta\omega} \int_{-\omega_2}^{-\omega_1} \cos \omega\tau \, d\omega + \frac{a^2}{4\Delta\omega} \int_{\omega_1}^{\omega_2} \cos \omega\tau \, d\omega \qquad \text{(b)}$$

$$= \frac{a^2}{2\Delta\omega} \left[\frac{1}{\tau} \sin \omega\tau \right]_{\omega_1}^{\omega_2} \qquad \text{(c)}$$

$$= \frac{a^2}{\tau\Delta\omega} \cos \left[\left(\frac{\omega_1 + \omega_2}{2} \right)\tau \right] \sin \left[\left(\frac{\omega_2 - \omega_1}{2} \right)\tau \right] \qquad \text{(d)}$$

This function is plotted in Figure 2.16.

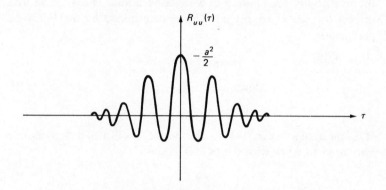

Figure 2.16   Autocorrelation function for narrow-band random process

A typical variable with an autocorrelation function like (d) is plotted in Figure 2.17. For a narrow-band process, the record will take the form of a modulated sine wave with no negative maxima, or positive minima.

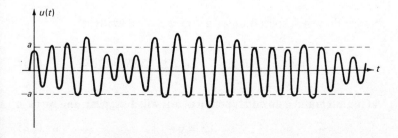

Figure 2.17   Sample record of narrow-band process; $\bar{\omega} = (\omega_1 + \omega_2)/2$, period $\simeq 2\pi/\omega$

Note. The broad-band process has a narrower range of τ for which $R_{uu}(\tau)$ is an appreciable function. The variance is then given by:

$$R_{uu}(0) = \frac{a^2}{2} \tag{e}$$

2.7 CROSSING ANALYSIS

In order to study the distribution of peak values, one needs to know the probability distribution of a function u and its derivative with respect to time $\dot{u}$, i.e. $p(u, \dot{u})$. This is determined by the following parameters:

$$\text{means} \quad U, \dot{U}$$

$$\text{deviations} \quad \sigma_{uu}, \sigma_{\dot{u}\dot{u}} \tag{2.81}$$

$$\text{correlation} \quad \rho_{u\dot{u}}$$

If the mean of u is equal to zero, the mean of $\dot{u}$ will also be zero and we only need to work with the two variances:

$$\sigma_{uu}^2 = \int_{-\infty}^{\infty} S_{uu}(\omega)\, d\omega$$

$$\sigma_{\dot{u}\dot{u}}^2 = \int_{-\infty}^{\infty} \omega^2 S_{uu}(\omega)\, d\omega \tag{2.82}$$

and the normalised covariance is:

$$\rho_{u\dot{u}} = \frac{\langle u\dot{u} \rangle}{\sigma_u \sigma_{\dot{u}}} \tag{2.83}$$

Notice that the expectation of u and $\dot{u}$ can be written:

$$\langle u\dot{u} \rangle = \frac{d}{d\tau} R_{uu}(\tau) \bigg|_{\tau=0} = i \int_{-\infty}^{\infty} \omega S_{uu}(\omega)\, d\omega \tag{2.84}$$

As the integrand is an odd function of ω it will disappear, and we have:

$$\rho_{u\dot{u}} \equiv 0 \tag{2.85}$$

This shows that, for a stationary process, u and its derivative $\dot{u}$ are uncorrelated; hence the probability distribution is:

$$p(u, \dot{u}) = \frac{1}{2\pi\sigma_{uu}\sigma_{\dot{u}\dot{u}}} \exp\left[-\tfrac{1}{2}\left(\frac{u^2}{\sigma_{uu}^2} + \frac{\dot{u}^2}{\sigma_{\dot{u}\dot{u}}^2}\right)\right]$$

$$= p(u)\,p(\dot{u}) \tag{2.86}$$

The curve is shown in Figure 2.18.

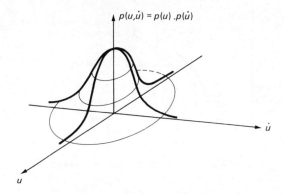

Figure 2.18   Probability distribution of u and $\dot{u}$

Let us now go back to the original problem of finding the number of crossings in a process u (Figure 2.19). The question is how many cycles have amplitudes larger than $u = a$ during a period T, i.e. how many cycles have *positive slopes* at $u = a$ during T. If the process is stationary the expected number of crossings per unit time is always

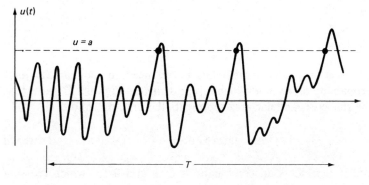

Figure 2.19   Positive-slope crossings, at u = a, for a random record

the same, and the total number of positive-slope crossings N_a^+ will only be a function of the period T and the average frequency of positive-slope crossings at $u = a$, v_a^+. We now intend to deduce this v_a^+ value by considering a dt increment (Figure 2.20) and assuming that the process is *narrow-band* (this implies that u is a smooth function without sudden changes of direction).

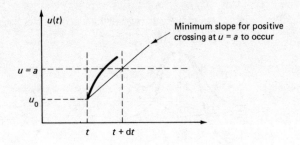

Figure 2.20   *Slope for a positive crossing*

For a crossing to exist during a particular dt, the minimum slope of u at t has to be:

$$\frac{a - u_0}{dt} \tag{2.87}$$

Hence the conditions for a positive slope to exist during dt are:

$$u_0 < a \tag{2.88}$$

and

$$\dot{u} > \frac{a - u_0}{dt} \tag{2.89}$$

To ascertain how possible this is, we need to look at the joint probability of u and $\dot{u}$, which is represented in Figure 2.21. Note that the shaded triangular area represents the limiting condition:

$$\tan \alpha = dt = \frac{a - u_0}{\dot{u}} \tag{2.90}$$

and it represents all combinations of u and $\dot{u}$ for which crossing occurs.

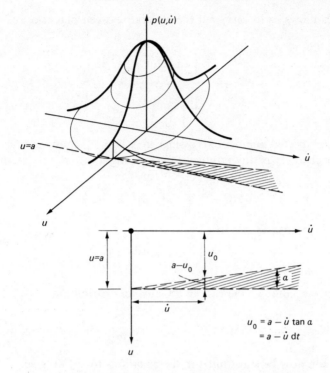

Figure 2.21   Probability of a positive slope occurring

The probability that the values of u and $\dot{u}$ will be in the triangular area is:

$$dP = \text{Prob} \,(+\text{ve slope crossing } u = a \text{ in time } dt)$$

$$= \int_0^\infty \left[\int_{u_0}^a p(u, \dot{u}) \, du \right] d\dot{u} \tag{2.91}$$

If $dt \to 0$, we can write:

$$p(u, \dot{u}) \Rightarrow p(a, \dot{u}) = \text{joint probability density } p(u, \dot{u}) \text{ evaluated at } u = a \tag{2.92}$$

and:

$$dP = \int_0^\infty \left[p(a, \dot{u}) \, du \right] d\dot{u} \tag{2.93}$$

which allows us to carry out the integral between brackets, i.e.

$$dP = \int_0^\infty [p(a, \dot{u})\dot{u}\,dt]\,d\dot{u}$$

$$= dt \left[\int_0^\infty p(a, \dot{u})\dot{u}\,d\dot{u} \right] \tag{2.94}$$

The expected number of crossings in dt is $= v_a^+\,dt$ and is equal to the probability of any positive slope crossing $u = a$ in time dt; thus:

$$v_a^+\,dt = \left[\int_0^\infty p(a, \dot{u})\dot{u}\,d\dot{u} \right] dt \tag{2.95}$$

$$v_a^+ = \int_0^\infty p(a, \dot{u})\dot{u}\,d\dot{u}$$

If we now consider the case of a Gaussian distribution:

$$p(a, \dot{u}) = \frac{1}{\sqrt{(2\pi)}\sigma_{uu}} \exp(-a^2/2\sigma_{uu}^2) \frac{1}{\sqrt{(2\pi)}\sigma_{\dot{u}\dot{u}}} \exp(-\dot{u}^2/2\sigma_{\dot{u}\dot{u}}^2) \tag{2.96}$$

This can now be substituted in the equations for v_a^+ to give:

$$v_a^+ = \frac{1}{2\pi\sigma_{uu}} \exp(-a^2/2\sigma_{uu}^2) \frac{1}{\sigma_{\dot{u}\dot{u}}} \int_0^\infty \dot{u} \exp(-\dot{u}^2/2\sigma_{\dot{u}\dot{u}}^2)\,d\dot{u} \tag{2.97}$$

(Notice that this type of integral has already been given in the footnote on page 18.) The value of v_a^+ can now be written:

$$v_a^+ = \frac{1}{2\pi} \frac{\sigma_{\dot{u}\dot{u}}}{\sigma_{uu}} \exp(-a^2/2\sigma_{uu}^2) \tag{2.98}$$

When $a = 0$ we have the 'zero crossing' frequency of the process:

$$v_0^+ = \frac{\sigma_{\dot{u}\dot{u}}}{2\pi\sigma_{uu}} \tag{2.99}$$

In terms of spectral densities this zero crossing frequency is:

$$v_0^+ = \frac{\sigma_{\dot{u}\dot{u}}}{2\pi\sigma_{uu}} = \frac{1}{2\pi} \left[\frac{\int \omega^2 S_{uu}(\omega)\,d\omega}{\int S_{uu}(\omega)\,d\omega} \right]^{\frac{1}{2}} = \frac{1}{2\pi} \left[\frac{M_2}{M_0} \right]^{\frac{1}{2}} \tag{2.100}$$

For a narrow-band process of characteristic frequency ω_0 we may write:

$$v_0^+ = \frac{\omega_0}{2\pi} \qquad (2.101)$$

2.8 PEAK DISTRIBUTION

Assume that we now want to find the peaks of a function between a and $a + da$ values of u (Figure 2.22). This probability is:

$$p_p(a)\,da \qquad (2.102)$$

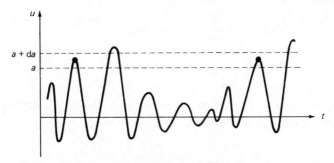

Figure 2.22   Peaks in the band a to a + da

The probability of peaks exceeding a is:

$$\int_a^\infty p_p(a)\,da \qquad (2.103)$$

For a narrow-band process there will be $v_0^+ T$ cycles in time T, with only $v_a^+ T$ cycles exceeding $u = a$. Hence the proportion of cycles with peaks exceeding a is v_a^+/v_0^+. This is the probability of peaks exceeding a:

$$\int_a^\infty p_p(a)\,da = \frac{v_a^+}{v_0^+} \qquad (2.104)$$

Differentiating with respect to a:

$$-p_p(a) = \frac{1}{v_0^+}\frac{d}{da}(v_a^+) \qquad (2.105)$$

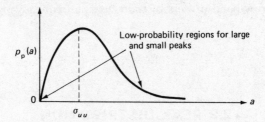

Figure 2.23 Rayleigh distribution

This result applies for any narrow-band process. If in addition u is a Gaussian function we obtain:

$$p_p(a) = -\frac{d}{da} \exp(-a^2/2\sigma_{uu}^2)$$

$$= \frac{a}{\sigma_{uu}^2} \exp(-a^2/2\sigma_{uu}^2) \qquad (0 \leqslant a \leqslant \infty) \qquad (2.106)$$

This probability function corresponds to a *Rayleigh* distribution (Figure 2.23). It has a maximum value at $a = \sigma_{uu}$.

2.9 MAXIMA OF A GENERAL PROCESS

Up to now we have assumed that the process under study was Gaussian and narrow-band. Let us now check if the narrow-band assumption is valid.

In this case a maximum will be defined by:

$$\dot{u} = 0 \quad \text{and} \quad \ddot{u} \text{ negative}$$

If the process is narrow-band, the frequency of a zero crossing v_0^+ and maximum v_m are the same. The frequency of $\dot{u} = 0$, which we call $v_{\dot{u}=0}^+$, is also the same as v_m.

$$v_m = v_{\dot{u}=0}^+$$

We can calculate v_m by substituting $\dot{u}$ for u and $\ddot{u}$ for $\dot{u}$ in the formula previously obtained for v_a^+. Putting $\dot{u} = a = 0$, we obtain:

$$v_m = \frac{1}{2\pi} \frac{\sigma_{\ddot{u}\ddot{u}}}{\sigma_{\dot{u}\dot{u}}} \qquad (2.107)$$

For any process the frequency of maxima can be given as:

$$v_{\mathrm{m}} = \frac{1}{2\pi} \frac{\sigma_{\ddot{u}\ddot{u}}}{\sigma_{\dot{u}\dot{u}}} = \frac{1}{2\pi} \left[\frac{\displaystyle\int_{-\infty}^{\infty} \omega^4 S_{uu}(\omega)\mathrm{d}\omega}{\displaystyle\int_{-\infty}^{\infty} \omega^2 S_{uu}(\omega)\mathrm{d}\omega} \right]^{\frac{1}{2}} \tag{2.108}$$

which can be compared with the statistical zero crossing frequency.

$$v_0^+ = \frac{1}{2\pi} \frac{\sigma_{\dot{u}\dot{u}}}{\sigma_{uu}} = \frac{1}{2\pi} \left(\frac{M_2}{M_0} \right)^{\frac{1}{2}} \tag{2.109}$$

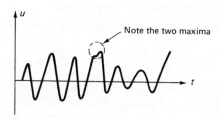

Figure 2.24   Random process

If they are not the same, the process is not narrow-band. The reason for the discrepancy is that in a non-narrow process we cannot assume that u varies as a sinusoidal function of variable amplitude, and components of different frequencies may produce additional maxima, as shown in Figure 2.24.

Determination of ϵ

The parameter ε defines the relative spectral width of a system, and can be determined from the proportion of negative maxima. In the case of interest to us this corresponds to the proportion of wave crests, which are in fact below the still water level. This proportion is:

$$r = \tfrac{1}{2}\left(1 - \frac{v_0^+}{v_{\mathrm{m}}^+} \right) \tag{2.110}$$

From previous equations:

$$v_m = \frac{1}{2\pi} \left(\frac{M_4}{M_2} \right)^{\frac{1}{2}} = \frac{1}{T_m} \qquad v_0^+ = \frac{1}{2\pi} \left(\frac{M_2}{M_0} \right)^{\frac{1}{2}} = \frac{1}{T_0^+} \quad (2.111)$$

Hence:

$$r = \tfrac{1}{2} \left[1 - \frac{M_2}{(M_0 M_4)^{\frac{1}{2}}} \right] = \tfrac{1}{2} \left[1 - (1 - \varepsilon^2)^{\frac{1}{2}} \right] \qquad (2.112)$$

Conversely:

$$\varepsilon^2 = 1 - (1 - 2r)^2 \qquad (2.113)$$

In this way, to calculate ε one needs simply to count the number of positive and negative maxima in the record.

Example 2.4

As an example of a broad-band process we shall consider the spectral density function for the surface elevation of a fully developed sea. For simplicity we shall consider the spectrum corresponding to this sea state to be the truncated white-noise spectrum illustrated in Figure 2.25. The autocorrelation function corresponding to this spectrum will be given by:

$$R_{\eta\eta}(\tau) = \frac{a^2}{\tau \Delta \omega} \cos \left[\left(\frac{\omega_1 + \omega_2}{2} \right) \tau \right] \sin \left[\left(\frac{\omega_2 - \omega_1}{2} \right) \tau \right] \qquad (a)$$

(See Example 2.3.) Here ω_1 and ω_2 are widely separated (typically we may take $\omega_1 = 0.3$ and $\omega_2 = 2.5$), and the function $R_{\eta\eta}$ is now sharply peaked around $\tau = 0$ and appreciable only near $\tau = 0$.

We shall see in Chapter 3 that, for a linear wave of frequency ω travelling in the positive x direction, the water particle velocity in this direction, at depth z, is given by:

$$v_x = \frac{g\kappa}{\omega} \frac{\cosh \left[\kappa(z + d) \right]}{\cosh \kappa d} \eta \qquad (b)$$

where z is measured from the still water level
$\quad d$ is the depth of the water
$\quad \kappa$ is the wavenumber, given by:

$$\omega^2 = g\kappa \tanh \kappa d \qquad (c)$$

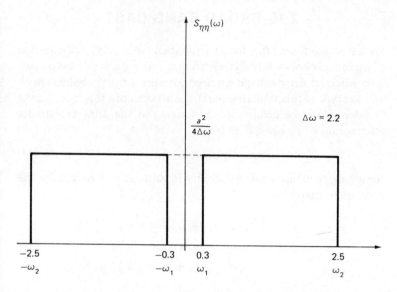

Figure 2.25   Spectral density of a (broad) band limited process

The horizontal water particle acceleration is given by:

$$\dot{v}_x = i\kappa g \, \frac{\cosh\left[\kappa(z+d)\right]}{\cosh \kappa d} \, \eta \tag{d}$$

Thus we may write down the cross spectral density of η and v_x as:

$$S_{\eta v_x} = \frac{g\kappa}{\omega} \, \frac{\cosh\left[\kappa(z+d)\right]}{\cosh \kappa d} \, S_{\eta\eta} \tag{e}$$

and η and $\dot{v}_x$:

$$S_{\eta \dot{v}_x} = i\kappa g \, \frac{\cosh\left[\kappa(z+d)\right]}{\cosh \kappa d} \, S_{\eta\eta} \tag{f}$$

and the cross spectral density of the water particle velocity and acceleration is given by:

$$S_{v_x \dot{v}_x} = \frac{ig^2\kappa}{\omega} \, \frac{\cosh^2\left[\kappa(z+d)\right]}{\cosh^2 \kappa d} \, S_{\eta\eta} \tag{g}$$

Note. $S_{\eta\eta}$ is real, $S_{\eta v_x}$ and $S_{v_x \dot{v}_x}$ have zero real part and are purely imaginary, as η and $\dot{v}_x$, and v_x and $\dot{v}_x$ are out of phase by $90°$.

2.10 BROAD-BAND CASE

We have already seen that for a narrow-band process the distribution of maxima follows a Rayleigh distribution. When the process cannot be considered narrow-band we need to express the probability with another type of function. In general, the maxima of a function u in the dt interval will be determined by looking at the joint probability distribution of u, $\dot{u}$ and $\ddot{u}$, i.e.

$$p(u, \dot{u}, \ddot{u}) \tag{2.114}$$

Assuming that this joint probability is gaussian, we may write the correlation matrix as:

$$\mu = \begin{bmatrix} \langle u, u \rangle & \langle u, \dot{u} \rangle & \langle u, \ddot{u} \rangle \\ \langle \dot{u}, u \rangle & \langle \dot{u}, \dot{u} \rangle & \langle \dot{u}, \ddot{u} \rangle \\ \langle \ddot{u}, u \rangle & \langle \ddot{u}, \dot{u} \rangle & \langle \ddot{u}, \ddot{u} \rangle \end{bmatrix} \tag{2.115}$$

Remembering the relationships between spectral densities and expectations, we have:

$$\mu = \begin{bmatrix} M_0 & 0 & -M_2 \\ 0 & M_2 & 0 \\ -M_2 & 0 & M_4 \end{bmatrix} \tag{2.116}$$

$$|\mu| = M_2 M_0 M_4 - M_2^3 \tag{2.117}$$

Hence:

$$p(u, \dot{u}, \ddot{u}) = \frac{1}{(2\pi)^{3/2} |\mu|^{1/2}} \times$$

$$\times \exp\left[-\frac{1}{2}\left(\frac{\dot{u}^2}{M_2} + \frac{M_4}{M_0 M_4 - M_2^2} u^2 + \right.\right.$$

$$\left.\left. + 2\frac{M_2}{M_0 M_4 - M_2^2} u\ddot{u} + \frac{M_0}{M_0 M_4 - M_2^2} \ddot{u}^2 \right) \right] \tag{2.118}$$

We can now compute the probability distribution of maxima in the dt interval. For a dt we have:

$$d\dot{u} \simeq |\ddot{u}|\, dt \qquad \ddot{u} \text{ negative} \tag{2.119}$$

The probability that u lies in the band u to $u + du$ (or a to $a + da$) and is near a maximum is:

$$\int_{-\infty}^{0} \left[p(u, \dot{u}, \ddot{u}) \, du \, d\dot{u} \right] d\ddot{u}$$

$$= \int_{-\infty}^{0} p(u, 0, \ddot{u}) \, du \, |\ddot{u}| \, dt \, d\ddot{u} \qquad (2.120)$$

$$= \left[\int_{-\infty}^{0} (p(u, 0, \ddot{u}) \, du \, |\ddot{u}|) \, d\ddot{u} \right] dt$$

This is equal to the mean frequency of maxima multiplied by dt in the u to $u + du$ interval, i.e.

$$v_{m}(u) \, du \, dt = \int_{-\infty}^{0} p(u, 0, \ddot{u}) |\ddot{u}| \, d\ddot{u} \, du \, dt \qquad (2.121)$$

The probability distribution of maxima is found by dividing this distribution by the total mean frequency of maxima, which is:

$$v_{mT} = \int_{-\infty}^{\infty} \left[\int_{-\infty}^{0} p(u, 0, \ddot{u}) |\ddot{u}| \, d\ddot{u} \right] du \qquad (2.122)$$

Hence:

$$p(u) = \frac{v_{m}}{v_{mT}} \qquad (2.123)$$

The function $v_{m}(u)$ can be computed knowing p and integrating. It is:

$$v_{m}(u) = \frac{1}{(2\pi)^{3/2} \sqrt{[M_{2}(M_{0}M_{4} - M_{2}^{2})]}} \times$$

$$\times \int_{-\infty}^{0} \exp \left[-\frac{1}{2} \left(\frac{M_{4}}{M_{0}M_{4} - M_{2}^{2}} u^{2} + 2 \frac{M_{2}}{M_{0}M_{4} - M_{2}^{2}} u\ddot{u} + \right.\right.$$

$$\left.\left. + \frac{M_{0}}{M_{0}M_{4} - M_{2}^{2}} \ddot{u}^{2} \right) \right] |\ddot{u}| \, d\ddot{u} \qquad (2.124)$$

This integral can be evaluated, giving:

$$v_{m}(u) = \frac{\sqrt{(M_{0}M_{4} - M_{2}^{2})}}{(2\pi)^{3/2} M_{0} \sqrt{M_{2}}} \exp \left(-\frac{1}{2} \frac{u^{2}}{M_{0}} \right) \times$$

$$\times \left[\exp\left(-\frac{1}{2}\frac{u^2}{M_0}\frac{M_2^2}{(M_0M_4 - M_2^2)} \right) + \right.$$

$$+ \frac{u}{\sqrt{M_0}}\frac{M_2}{\sqrt{(M_0M_4 - M_2^2)}} \times$$

$$\left. \times \int_{-\frac{uM_2}{\sqrt{M_0}\sqrt{(M_0M_4 - M_2^2)}}}^{\infty} \exp(-\tfrac{1}{2}x^2)\,\mathrm{d}x \right] \qquad (2.125)$$

where x = dummy variable of integration.

We can write $v_m(u)$ as a function of the ε parameter, i.e.

$$\varepsilon = \left(1 - \frac{M_2^2}{M_0M_4} \right)^{\frac{1}{2}} \qquad (2.126)$$

$$v_m(u) = \frac{\varepsilon\sqrt{M_4}}{(2\pi)^{3/2}\sqrt{(M_0M_2)}}\exp\left(-\frac{1}{2}\frac{u^2}{M_0} \right)\left[\exp\left(-\frac{1}{2}\frac{u^2}{M_0^2}\frac{M_2^2}{M_4}\frac{1}{\varepsilon^2} \right) + \right.$$

$$\left. + \frac{u}{M_0\sqrt{M_4}}\frac{M_2}{\varepsilon}\int_{-\frac{uM_2}{M_0\sqrt{M_4}}\frac{1}{\varepsilon}}^{\infty} \exp(-\tfrac{1}{2}x^2)\,\mathrm{d}x \right]$$

$$= \frac{\sqrt{M_4}}{(2\pi)^{3/2}\sqrt{(M_0M_2)}}\left[\varepsilon\exp\left(-\frac{1}{2}\frac{u^2}{M_0}\frac{1}{\varepsilon^2} \right) + \right.$$

$$\left. + (1 - \varepsilon^2)^{\frac{1}{2}}\frac{u}{\sqrt{M_0}}\int_{-\infty}^{\frac{u}{\sqrt{M_0}}\frac{\sqrt{(1-\varepsilon^2)}}{\varepsilon}} \exp(\tfrac{1}{2}x^2)\,\mathrm{d}x \right] \qquad (2.127)$$

Now the mean frequency of all maxima is:

$$v_{mT} = \frac{1}{2\pi}\left(\frac{M_4}{M_2} \right)^{\frac{1}{2}} \qquad (2.128)$$

We can now calculate the probability:

$$p(u) = \frac{v_m}{v_{mT}} \qquad (2.129)$$

$$p(u) = \frac{1}{(2\pi)^{1/2}\sqrt{M_0}} \left[\varepsilon \exp\left(-\frac{1}{2}\frac{u^2}{M_0}\frac{1}{\varepsilon^2} \right) + (1-\varepsilon^2)^{\frac{1}{2}} \times \right.$$

$$\left. \times \frac{u}{\sqrt{M_0}}\exp\left(-\frac{1}{2}\frac{u^2}{M_0} \right) \int_{-\infty}^{\frac{u}{\sqrt{M_0}}\frac{\sqrt{(1-\varepsilon^2)}}{\varepsilon}} \exp\left(-\tfrac{1}{2}x^2 \right)dx \right] \quad (2.130)$$

It is possible to show that ε is always positive, varying between 0 and 1. When $\varepsilon \to 0$ we have an infinitely narrow spectrum:

$$p(u) = \frac{u}{M_0}\exp\left(-\frac{1}{2}\frac{u^2}{M_0} \right)$$

$$= \frac{u}{\sigma_u^2}\exp\left(-\frac{1}{2}\frac{u^2}{\sigma_u^2} \right) = \text{Rayleigh distribution (for } u \geqslant 0) \quad (2.131)$$

The other limiting case is when $\varepsilon \to 1$, for which:

$$p(u) = \frac{1}{\sqrt{(2\pi)}\sqrt{M_0}}\exp\left(-\frac{1}{2}\frac{u^2}{M_0} \right) \quad (2.132)$$

which is a Gaussian distribution. Graphs of $p(u)$ for these and intermediate values are shown in Figure 2.26.

The cumulative probability $P(u)$ may be defined as the probability of u exceeding a given u value.

$$P(u) = \int_u^\infty p(u)\,du \quad (2.133)$$

Substituting $p(u)$, we obtain the graphs shown in Figure 2.27.

In some cases it is convenient to consider the highest $1/n$th of the total number of maxima. The $1/n$th highest maxima correspond to those values of u greater than u', say, where:

$$P(u') = \int_{u'}^\infty p(u)\,du = \frac{1}{n} \quad (2.134)$$

The average value of u for these maxima can be called $U^{(1/n)}$, where:

$$U^{(1/n)} = \int_{u'}^\infty p(u)\,u\,du \quad (2.135)$$

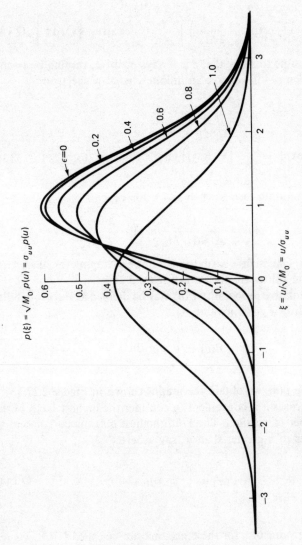

$p(\xi) = \sqrt{M_0}\,p(u) = \sigma_{uu}p(u)$

$\xi = u/\sqrt{M_0} = u/\sigma_{uu}$

Figure 2.26 Graphs of $p(\xi)$, the probability distribution of the heights of maxima ($\xi = u/\sqrt{M_0}$) for different values of the width ε of the energy spectrum (from Cartwright and Longuet-Higgins[1])

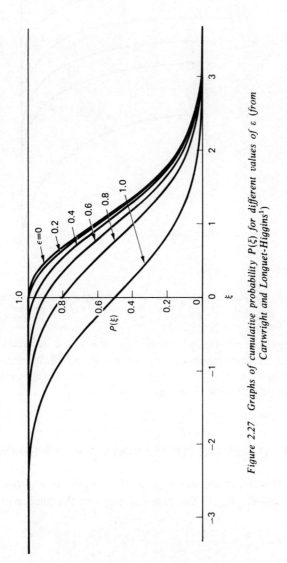

Figure 2.27 Graphs of cumulative probability $P(\xi)$ for different values of ε (from Cartwright and Longuet-Higgins[1])

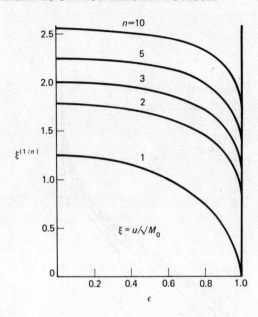

Figure 2.28 Graphs of $\xi^{(1/n)}$, mean height of the 1/nth highest maxima, as a function of ε, for n = 1, 2, 3, 5, 10 (from Cartwright and Longuet-Higgins[1])

$U^{(1/n)}$ has been computed[1] numerically for $n = 1, 2, 3, 5$ and 10 and for different ε, and the results are plotted in Figure 2.28. When $\varepsilon < 0.5$ the dependence of $U^{(1/n)}$ on ε is slight, but when ε approaches 1 the gradient of $\partial U^{(1/n)}/\partial \varepsilon$ tends to $-\infty$.

2.11 ANALYSIS OF MAXIMA, FINITE SAMPLES

We want to know the mean value of a sample of N maxima, which will be called $\langle u_{\max} \rangle$. The proof is long but the final result[1] is:

$$\langle u_{\max} \rangle \simeq \sqrt{2} \{ [\ln(1 - \varepsilon^2)^{\frac{1}{2}} N]^{\frac{1}{2}} + \tfrac{1}{2} \gamma [\ln(1 - \varepsilon^2)^{\frac{1}{2}} N]^{-\frac{1}{2}} \} \sqrt{M_0}$$

(2.136)

where $\gamma =$ Euler's constant $= 0.5772 \ldots$

When $\varepsilon \to 0$ the equation reduces to:

$$\langle u_{max} \rangle \simeq \sqrt{2}\left(\sqrt{\ln N} + \frac{1}{2}\gamma\,\frac{1}{\sqrt{\ln N}}\right)\sqrt{M_0} \qquad (2.137)$$

When $\varepsilon \to 1$ the above formula is no longer valid, but it is expected to cover all practical applications.

It is also shown that the most probable value of the maximum or mode of u_{max} is:

$$(u_{max})_{mode} = \sqrt{M_0}\left[\sqrt{(2\ln N)}\right] \qquad (2.138)$$

Example 2.5

Using the band-limited white noise spectrum mentioned in Example 2.4 we are now in a position to calculate the frequency of zero crossings, v_0^+, and the mean zero crossing period for the process as well as the frequency of maxima, v_m. The mean zero crossing period is an important parameter for the statistical analysis of sea states.

We start by calculating the moments of the spectrum. We have:

$$M_0 = \int_{-\infty}^{\infty} S_{\eta\eta}(\omega)\,d\omega = \frac{2a^2}{4(\omega_2 - \omega_1)}\int_{\omega_1}^{\omega_2} d\omega = \frac{a^2}{2} \qquad (a)$$

$$M_2 = \int_{-\infty}^{\infty} \omega^2 S_{\eta\eta}(\omega)\,d\omega = \frac{2a^2}{4(\omega_2 - \omega_1)}\int_{\omega_1}^{\omega_2} \omega^2\,d\omega = \frac{a^2(\omega_2^3 - \omega_1^3)}{6(\omega_2 - \omega_1)} \qquad (b)$$

$$M_4 = \int_{-\infty}^{\infty} \omega^4 S_{\eta\eta}(\omega)\,d\omega = \frac{2a^2}{4(\omega_2 - \omega_1)}\int_{\omega_1}^{\omega_2} \omega^4\,d\omega = \frac{a^2(\omega_2^5 - \omega_1^5)}{10(\omega_2 - \omega_1)} \qquad (c)$$

Hence we obtain:

$$v_0^+ = \frac{1}{2\pi}\left(\frac{\omega_2^3 - \omega_1^3}{3(\omega_2 - \omega_1)}\right)^{\frac{1}{2}} \qquad (d)$$

and the mean zero crossing period $1/v_0^+$ is:

$$T_0 = 2\pi\left(\frac{3(\omega_2 - \omega_1)}{\omega_2^3 - \omega_1^3}\right)^{\frac{1}{2}} \qquad (e)$$

For a fully developed sea we obtain, with $\omega_1 = 0.3$ rad/s and $\omega_2 = 2.5$ rad/s:

$$T_0 \simeq 4.084 \text{ seconds} \qquad (f)$$

We may now obtain the frequency of maxima as:

$$v_m = \frac{1}{2\pi} \left(\frac{M_4}{M_2} \right)^{\frac{1}{2}} = \frac{1}{2\pi} \left(\frac{3(\omega_2^5 - \omega_1^5)}{5(\omega_2^3 - \omega_1^3)} \right)^{\frac{1}{2}} \tag{g}$$

For the above values of ω_1 and ω_2:

$$v_0^+ = 0.245 \, s^{-1} \tag{h}$$

$$v_m = 0.308 \, s^{-1} \tag{i}$$

Note. For a narrow-band process we would have $v_0^+ = v_m$. The spectral width parameter for this spectrum is:

$$\varepsilon = \left(1 - \frac{M_2}{M_0 M_4} \right)^{\frac{1}{2}} = \left(1 - \frac{5}{9} \frac{(\omega_2^3 - \omega_1^3)^2}{(\omega_2 - \omega_1)(\omega_2^5 - \omega_1^5)} \right)^{\frac{1}{2}} \tag{j}$$

Hence: $\varepsilon \simeq 0.6$ \hfill (k)

Example 2.6

In order to perform a 'design wave analysis' of an offshore structure it is essential to know the significant wave height H_s and the mean zero crossing period T_0. The significant wave height is the mean of the highest third of the waves, and is discussed in Chapter 3.

For the band-limited spectrum with $\omega_1 = 0.3 \, rad/s$ and $\omega_2 = 2.5 \, rad/s$, we have already obtained the mean zero crossing period. We have:

$$T_0 = 4.084 \, s \tag{a}$$

We also calculated: \hfill $\varepsilon = 0.6$ \hfill (b)

For a narrow-band process we shall see that:

$$H_s \simeq 2 \times 2 \sqrt{M_0} \tag{c}$$

The first factor refers to the fact that the wave height for a narrow-band process is twice the maximum elevation; the second factor is obtained from the curve in Figure 2.28 corresponding to $n = 3$, which is the one for the significant wave as defined above. If we extend the use of formula (c) to our spectrum (as we may for all practical purposes), we obtain for $\varepsilon = 0.6$ that:

$$H_s = 2 \times 1.9 \sqrt{M_0} \tag{d}$$

This gives, as the significant wave height for our spectrum:

$$H_s = 2.687a \qquad \text{(e)}$$

2.12 WEIBULL DISTRIBUTION

The Rayleigh distribution cannot always adequately represent certain maxima processes. This distribution is a one parameter distribution fully defined by knowing the value of the standard deviation. In practice many processes are better defined by a more flexible two-parameter distribution called the Weibull distribution. Its probability density function can be written as:

$$p(u) = \frac{K}{c} \left(\frac{u}{c}\right)^{K-1} \exp\left[-\left(\frac{u}{c}\right)^K\right] \qquad (2.139)$$

with a cumulative probability given by:

$$P(> u) = \exp\left[-\left(\frac{u}{c}\right)^K\right] \qquad (2.140)$$

Note that the Rayleigh distribution is a special case of Weibull's for $K = 2$ and $\sigma = c/\sqrt{2}$.

The coefficients K and c can be obtained by plotting the results on a double logarithmic paper called Weibull's paper and with the scale obtained by taking double logarithms of equation (2.139), i.e.

$$\frac{1}{K} \ln\left\{-\ln\left[P(> u)\right]\right\} = \ln u - \ln c \qquad (2.141)$$

The results should now fit a line with slope $1/K$, and c can be determined by intersecting the line of the results with $\ln\left\{-\ln\left[P(> u)\right]\right\} = 0$.

The Weibull distribution is frequently used to plot wave heights for different sea states. It is also used to describe the distribution of wind speed and wind direction, for which case the K and c coefficients are taken to be functions of the compass direction.

Note that for $K = 1$ the Weibull gives the exponential distribution, with:

$$p(u) = \frac{1}{c} \exp\left(-\frac{u}{c}\right)$$

and
$$P(u) = \exp\left(-\frac{u}{c}\right)$$
(2.142)

with only one parameter to be determined.

Reference

1. Cartwright, D. E., and Longuet–Higgins, M. S., 'The statistical distribution of the maxima of a random function', *Proc. R. Soc. (Series A)*, **237** (1956)

Bibliography

Newland, D. E., *An introduction to random vibrations and spectral analysis*, Longman (1975)

Bendat, J. S., and Piersol, A. G., *Random data: analysis and measurement procedures*, John Wiley (1971)

Davenport, W. B., and Poot, W. L., *An introduction to the theory of random signals and noise*, McGraw-Hill (1958)

3 Sea Waves and Sea States

3.1 INTRODUCTION

In this chapter the more important concepts relating to sea waves and sea states will be reviewed. The aim is to provide the reader with the necessary background to enable him to interpret the oceanographic information in order to carry out the relevant structural analysis.

Two different methods of analysis, requiring different types of information, are currently used for offshore structures. The first is the design wave approach, and consists of assuming a wave of a given period and height which represents the maximum wave occurring for certain environmental conditions. This analysis requires only two parameters and is fully deterministic. The second approach is to work with the wave-energy spectrum using probabilistic theory, which allows us to obtain results for maximal stresses, displacements, etc., of the structure within a certain confidence level.

The latter approach is to be preferred, since the design wave does not give a sufficiently precise indication of the maximum response of the structure, which depends on how the structure reacts to the loading (e.g. a wave of the same or smaller height as the design wave but of longer period can produce a larger response if its period is nearer to resonance than the period of the design wave). If one uses a series of (design) waves of given heights and periods, as is usually done, the solutions thus obtained do not represent well the variability of the sea states. This is especially important for fatigue analysis, where a good history prediction of the stress magnitude and number of cycles is needed. It is more accurate to define the sea state as a wave-energy spectrum and analyse the structure using random vibrations. The use of random vibration theory generally assumes a linear analysis, and its relevance depends on the non-linear effects being susceptible to linearisation. One is then restricted to using a linear wave theory and compelled to linearise other non-linear terms such as drag.

After a brief introduction to the basic principles of wave dynamics, we will present and discuss the errors involved in using linear wave theory. Having (we trust) satisfied the reader with the validity of our approximation, we will then proceed to review the statistical procedures for analysing wave data and how the wave spectra are determined. The two more commonly used spectra, Pierson–Moskowitz and JONSWAP, are then discussed in detail and a way of making them directional is described.

3.2 SURFACE WAVES

Basic concepts

Here the inviscid motion of an incompressible fluid of finite depth d will be considered. One takes coordinates (x, y, z) fixed in space, where z is the vertical coordinate increasing upwards, and with the origin at the still water level, SWL (Figure 3.1). An Eulerian velocity vector can be associated with each point in space.

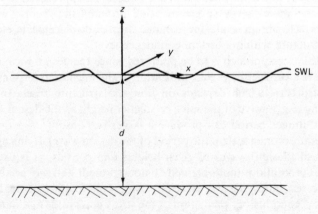

Figure 3.1   Coordinate system

$$\vec{v} = (v_x, v_y, v_z)$$

The v_x, v_y, v_z are the velocity components in the direction of the axes. The assumption that the fluid is incompressible and inviscid is reasonable for the ocean and will be used here when describing the kinematics of waves. We shall, however, calculate the drag forces on members of steel offshore structures; this is not done by considering the full viscous flow equations, but by the use of an empirical formula (Morison's equation) that is applicable for typical members of steel platforms. In calculating the forces on large-diameter members such as the columns of gravity platforms, a full diffraction theory is used; but (as will be seen later) the inertia forces dominate the drag forces, so we can use the above assumptions as to the nature of the fluid motion in this case also.

Consider now the Navier–Stokes equations for the motion of an incompressible fluid, i.e.

$$\rho \frac{D\vec{v}}{Dt} = \rho\vec{F} - \vec{\nabla}p - \mu\vec{\nabla} \times \vec{\omega} \tag{3.1}$$

where ρ is the density of the fluid

$\dfrac{D}{Dt}$ is the material derivative, defined by:

$$\frac{D}{Dt} = \frac{\partial}{\partial t} + \vec{v} \cdot \vec{\nabla} \tag{3.2}$$

$\vec{\nabla}$ is the gradient vector $(\partial/\partial x, \partial/\partial y, \partial/\partial z)$
p is the pressure
$\vec{F}$ is the body force vector
μ is the viscosity
$\vec{\omega}$ is the vorticity field defined by the curl of the velocity vector, i.e.

$$\vec{\omega} = \vec{\nabla} \times \vec{v} \tag{3.3}$$

Note that $\vec{\omega}$ is a measure of the rotationality of the fluid. From the equation one can see that the assumption of zero viscosity *has the same effect* on the equations of motion as the assumption that $\vec{\omega} = \vec{0}$ throughout the fluid, or:

$$\vec{\nabla} \times \vec{v} = \vec{0} \tag{3.4}$$

Equation (3.4) is called the irrotationality condition. A logical consequence of this condition is the existence of a velocity potential $\Phi(x, y, z)$ defined by:

$$\vec{v} = \vec{\nabla}\Phi = \left(\frac{\partial\Phi}{\partial x}, \frac{\partial\Phi}{\partial y}, \frac{\partial\Phi}{\partial z}\right) \tag{3.5}$$

Surfaces of constant Φ are surfaces of constant potential energy. In steady flow they are perpendicular to the direction of flow of the fluid.

In addition to the equations describing conservation of momentum, mass conservation requires that:

$$\frac{D\rho}{Dt} + \rho\vec{\nabla} \cdot \vec{v} = 0 \tag{3.6}$$

which is called the continuity equation. Incompressibility of a fluid implies that the density remains constant, hence:

$$\frac{D\rho}{Dt} = 0 \tag{3.7}$$

Consequently:

$$\vec{\nabla} \cdot \vec{v} = 0 \tag{3.8}$$

for an incompressible homogeneous fluid. Substituting the velocity vector in terms of the potential Φ, the incompressible inviscid fluid is governed by:

$$\vec{\nabla} \cdot \vec{\nabla}\Phi = \nabla^2\Phi = 0 \tag{3.9}$$

where

$$\nabla^2 = \frac{\partial^2}{\partial x^2} + \frac{\partial^2}{\partial y^2} + \frac{\partial^2}{\partial z^2}$$

Another formulation can also be used for *two-dimensional flow* fields. It is based on the stream function ψ defined by:

$$v_x = \frac{\partial\psi}{\partial y} \quad \text{and} \quad v_y = -\frac{\partial\psi}{\partial x} \tag{3.10}$$

The incompressibility equation ($\vec{\nabla} \cdot \vec{v} = 0$) is now satisfied identically, assuming that the function ψ is continuous. If one now considers the irrotationality condition,

$$\vec{\nabla} \times \vec{v} = \vec{\nabla} \times \left(\frac{\partial\psi}{\partial y}, -\frac{\partial\psi}{\partial x}, 0\right)$$

then
$$\vec{\nabla} \times \vec{v} = \left(0, 0, \frac{\partial^2 \psi}{\partial x^2} + \frac{\partial^2 \psi}{\partial y^2}\right) = \vec{0} \qquad (3.11)$$

or simply:

$$\nabla^2 \psi = 0 \qquad (3.12)$$

where ∇^2 is now the two-dimensional operator

$$\nabla^2 = \frac{\partial^2}{\partial x^2} + \frac{\partial^2}{\partial y^2}$$

Surfaces of ψ constant are the paths of the fluid particles in steady flow. Boundaries of solid surfaces in this formulation are surfaces of constant ψ.

Example 3.1

To illustrate the concepts presented above, consider steady parallel flow with a constant velocity V parallel to the x axis (Figure 3.2). One can write:

$$\vec{v} = (V, 0, 0) \qquad \text{(a)}$$

Then the *potential* is:

$$\Phi = Vx + C \qquad \text{(b)}$$

(where C is an arbitrary constant) and the surfaces of constant Φ are

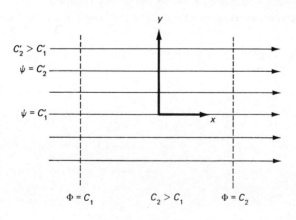

Figure 3.2   Steady parallel flow

$x = $ constant, the dotted lines in the figure representing the drop in potential energy of the particles between $x = (C_1 - C)/V$ and $x = (C_2 - C)/V$.

The *stream function* is given by:

$$\psi = Vy + C' \qquad \text{(c)}$$

(where C' is another arbitrary constant). Then surfaces of constant ψ are $y = $ constant, the solid lines in the figure representing for this steady flow the actual paths of the particles of the fluid.

For fully three-dimensional problems it is possible to define a vector stream function:

$$\vec{\psi} = (\psi_x, \psi_y, \psi_z) \qquad (3.13)$$

by:

$$\vec{v} = \vec{\nabla} \times \vec{\psi} \qquad (3.14)$$

such that (for uniqueness):

$$\vec{\nabla} \cdot \vec{\psi} = 0 \qquad (3.15)$$

The assumption of irrotationality and incompressibility results in the following three equations:

$$\nabla^2 \psi_x = 0 \qquad \nabla^2 \psi_y = 0 \qquad \nabla^2 \psi_z = 0 \qquad (3.16)$$

These equations are, however, difficult to use, because the application of boundary conditions now becomes very complex.

For axisymmetric flow with symmetry about the Z axis (Figure 3.3),

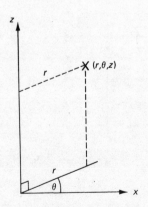

Figure 3.3   Cylindrical polar coordinates

one can define Stokes's stream function in cylindrical polar coordinates (r, θ, z) by:

$$v_r = \frac{1}{r}\frac{\partial \psi}{\partial z}$$

$$v_z = -\frac{1}{r}\frac{\partial \psi}{\partial r} \tag{3.17}$$

where v_r and v_z are the radial velocity and the velocity parallel to the Z axis, respectively. This formulation is suitable for geometries with a cylindrical solid surface and no θ dependence in the variables.

The conditions on the free surface

Consider a fluid on which disturbances of height $\eta(x, y, t)$ above still water level are propagating (Figure 3.4). The following conditions will occur on the free surface.

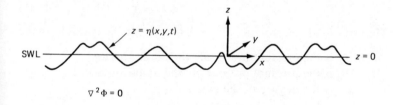

Figure 3.4   Boundary condition definitions

(a) Kinematic condition.   The vertical velocity at the free surface, taking into consideration that the surface moves with the fluid, is:

$$v_z = \frac{\partial \eta}{\partial t} + v_x \frac{\partial \eta}{\partial x} \qquad \text{at } z = \eta \tag{3.18}$$

If the slope $\partial \eta / \partial x$ or v_x is small we can neglect this term to obtain:

$$v_z \simeq \frac{\partial \eta}{\partial t} \qquad \text{at } z = \eta \qquad (3.19)$$

or, in terms of velocity potential:

$$\frac{\partial \eta}{\partial t} = \frac{\partial \Phi}{\partial z} \qquad \text{at } z = \eta \qquad (3.20)$$

Now by Taylor's theorem:

$$\Phi\bigg|_{z=\eta} = \Phi\bigg|_{z=0} + \eta \frac{\partial \Phi}{\partial z}\bigg|_{z=0} + 0(\eta^2) \qquad (3.21)$$

For small disturbances (small η or $\partial \Phi / \partial z$):

$$\frac{\partial \eta}{\partial t} \simeq \frac{\partial \Phi}{\partial z} \qquad \text{at } z = 0 \qquad (3.22)$$

(b) *Pressure condition.* We now assume that the surface is at a constant pressure (atmospheric pressure in the case of a water–air interface). So, from Bernoulli's equation (linearised) for irrotational motion, we have:

$$P_i - P_o = -\rho\left(\frac{\partial \Phi}{\partial t} + g\eta\right) \qquad (3.23)$$

where P_i is the pressure just inside the liquid surface
P_o is the pressure just outside the liquid surface
ρ is the density of the liquid
g is the acceleration due to gravity

If we neglect surface-tension effects, we have $P_i = P_o$.

$$\therefore \qquad \frac{\partial \Phi}{\partial t} = -g\eta \qquad \text{at } z = \eta \qquad (3.24)$$

or, by the argument above:

$$\frac{\partial \Phi}{\partial t} \simeq -g\eta \qquad \text{at } z = 0 \qquad (3.25)$$

Note that this equation gives the surface elevation when Φ is known. Combining the kinematic condition with the last equation, we obtain, after eliminating η:

$$\frac{\partial \Phi}{\partial z} \simeq -\frac{1}{g}\frac{\partial^2 \Phi}{\partial t^2} \qquad \text{at } z = 0 \qquad (3.26)$$

General solution for a fluid of constant depth

The third boundary condition of the problem is enforced at $z = -h$, assuming a solid impermeable surface here with negligible slope. We have no flux of liquid through this surface.

$$\therefore \qquad \frac{\partial \Phi}{\partial z} = 0 \qquad \text{at } z = -d \qquad (3.27)$$

If we now assume a separable solution:

$$\Phi(x, y, z, t) = \phi(x, y)f(z)\exp(i\omega t) \qquad (3.28)$$

Here we have used $\exp(i\omega t)$, which is defined by:

$$\exp(i\omega t) = \cos \omega t + i \sin \omega t \qquad (3.29)$$

The real part of $\exp(i\omega t)$ is written:

$$\text{Re}[\exp(i\omega t)] = \cos \omega t \qquad (3.30)$$

and the imaginary part:

$$\text{Im}[\exp(i\omega t)] = \sin \omega t \qquad (3.31)$$

For the separation of variables in the following, only the real part is needed. From the continuity equation:

$$\frac{\partial^2 \Phi}{\partial x^2} + \frac{\partial^2 \Phi}{\partial y^2} + \frac{\partial^2 \Phi}{\partial z^2} = 0 \qquad (3.32)$$

we obtain by substitution:

$$\nabla^2 \phi \, f(z) + \phi \, \frac{\partial^2 f}{\partial z^2} = 0 \qquad (3.33)$$

where now:

$$\nabla^2 = \frac{\partial^2}{\partial x^2} + \frac{\partial^2}{\partial y^2}$$

Hence we can write:

$$\frac{\nabla^2 \phi(x, y)}{\phi(x, y)} = -\frac{\partial^2 f/\partial z^2}{f(z)} \qquad (3.34)$$

For these functions to be equal for all x, y, z we must have:

$$\frac{\nabla^2 \phi}{\phi} = -\kappa^2$$

$$\frac{\partial^2 f/\partial z^2}{f} = \kappa^2 \tag{3.35}$$

where κ^2 is a constant to be determined by the boundary conditions.

The first of equations (3.35) gives the Helmholtz equation in two dimensions, i.e.

$$\nabla^2 \phi + \kappa^2 \phi = 0 \tag{3.36}$$

where $\phi(x, y)$ will be called the 'reduced velocity potential'. The second of equations (3.35) gives:

$$\frac{\partial^2 f}{\partial z^2} = \kappa^2 f \tag{3.37}$$

with the boundary condition, from equation (3.27):

$$\frac{\partial f}{\partial z} = 0 \qquad \text{for } \phi \neq 0 \quad \text{at } z = -d \tag{3.38}$$

and the boundary condition on the free surface—formula (3.22)—gives:

$$\frac{\partial f}{\partial z} = \frac{-\omega^2}{g} f \qquad \text{for } \phi \neq 0 \quad \text{at } z = 0 \tag{3.39}$$

From equation (3.37) with boundary conditions (3.38) and (3.39) the following result can be obtained:

$$f(z) = \frac{\cosh\left[\kappa(z+d)\right]}{\cosh \kappa d} \tag{3.40}$$

with $\omega^2 = g\kappa \tanh \kappa d$ the 'dispersion' relation for disturbances travelling on a liquid of depth d (g is the acceleration due to gravity). Hence:

$$\Phi(x, y, z, t) = \phi(x, y) \frac{\cosh\left[\kappa(z+d)\right]}{\cosh \kappa d} \exp(-i\omega t) \tag{3.41}$$

is the general solution for a harmonic disturbance where $\phi(x, y)$ is *any* solution of Helmholtz equation (3.36).

In the previous analysis we have been using a complex notation that often simplifies the calculations. The actual value of the physical variable is here assumed to be the real part of its complex counterpart. The structure of the complex number system makes sure that we do in fact get the right results by taking the real part after the calculation.

We can now calculate the wave amplitude using equation (3.25):

$$\eta(x, y) = +i\frac{\omega}{g}\,\phi(x, y) \tag{3.42}$$

Here the time-dependent factor $\exp(i\omega t)$ is assumed, and hence the i appearing in the right-hand side represents in complex notation the fact that ϕ is 90° out of phase with η, with respect to the angle ωt.

The complex notation can be seen at work here if we insert the time factor in the equation and take the real part of both sides.

$$\text{Re}\{\eta(x, y)\exp(i\omega t)\} = \text{Re}\{-i\frac{\omega}{g}\,\phi(x, y)\exp(i\omega t)\} \tag{3.43}$$

Note that we are assuming for the moment that η and ϕ are real. Whence:

$$\eta(x, y)\cos\omega t = \frac{\omega}{g}\,\phi(x, y)\sin\omega t$$
$$= \frac{\omega}{g}\,\phi(x, y)\cos(\omega t - 90°) \tag{3.44}$$

As $\phi(x, y)$ is a solution of the Helmholtz equation, so must be $\eta(x, y)$, as they are just proportional. (See equation (3.36).)

Linear wave theory

If we now consider a single sinusoidal wave of frequency ω (Figure 3.5) and take the x axis perpendicular to the wave crests and in the direction of propagation, we have a free surface disturbance of the form:

$$\eta(x, t) = \eta(x)\exp(i\omega t) \tag{3.45}$$

η is a solution of the one-dimensional Helmholtz equation:

$$\frac{\partial^2\eta}{\partial x^2} + \kappa^2\eta = 0 \tag{3.46}$$

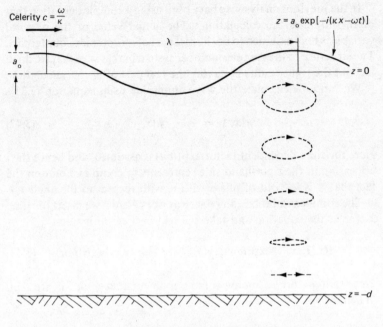

Figure 3.5   Linear or Airy wave

The elevation has a harmonic solution:

$$\eta(x) = a_0 \exp(-i\kappa x) \qquad (3.47)$$

where κ is the wavenumber, $\kappa = 2\pi/\lambda$. Whence:

$$\eta(x, t) = a_0 \exp[-i(\kappa x - \omega t)] \qquad (3.48)$$

This represents the elevation of a linear Airy wave of amplitude a_0 travelling in the direction of x and increasing with a phase velocity or celerity c, given by:

$$c = \omega/\kappa \qquad (3.49)$$

Utilising the linearised free-surface boundary conditions, we can write the velocity potential for the motion as:

$$\Phi(x, z, t) = i\frac{g}{\omega} a_0 \exp[-i(\kappa x - \omega t)] \frac{\cosh[\kappa(z+d)]}{\cosh \kappa d} \qquad (3.50)$$

and the 'dispersion' relation linking the frequency with the wavenumber, obtained by substituting this expression into the governing equation (3.9), is:

$$\omega^2 = g\kappa \tanh \kappa d \qquad (3.51)$$

The particle velocities in x and z directions are:

$$v_x = \frac{\partial \Phi}{\partial x} = \frac{g}{\omega} \kappa a_0 \exp[-i(\kappa x - \omega t)] \frac{\cosh[\kappa(z+d)]}{\cosh \kappa d} \qquad (3.52)$$

$$v_z = \frac{\partial \Phi}{\partial z} = -i \frac{g}{\omega} \kappa a_0 \exp[-i(\kappa x - \omega t)] \frac{\sinh[\kappa(z+d)]}{\cosh \kappa d} \qquad (3.53)$$

The particle trajectories, at mean depth $\bar{z}$ (assuming the fluid velocity in the displaced position to be the same as at $\bar{z}$), are:

$$r_x = \int_0^t v_x \, \mathrm{d}t = \frac{1}{i\omega} \frac{g\kappa}{\omega} a_0 \exp[-i(\kappa x - \omega t)] \frac{\cosh[\kappa(\bar{z}+d)]}{\cosh \kappa d} \qquad (3.54)$$

or, taking the real part for the actual displacement:

$$r_x = +a_0 \frac{g\kappa}{\omega^2} \sin(\kappa x - \omega t) \frac{\cosh[\kappa(\bar{z}+d)]}{\cosh \kappa d} \qquad (3.55)$$

$$r_y = \int_0^t v_y \, \mathrm{d}t = \frac{1}{i\omega} \frac{-ig\kappa}{\omega} a_0 \exp[-i(\kappa x - \omega t)] \frac{\sinh[\kappa(\bar{z}+d)]}{\cosh \kappa d} \qquad (3.56)$$

Taking the real part we obtain:

$$r_y = a_0 \frac{g\kappa}{\omega^2} \cos(\kappa x - \omega t) \frac{\sinh[\kappa(\bar{z}+d)]}{\cosh \kappa d} \qquad (3.57)$$

The particles describe ellipses centred on their mean position $z = \bar{z}$, with semi-axes (see Figure 3.5):

$$\frac{g\kappa}{\omega^2} a_0 \frac{\cosh[\kappa(\bar{z}+d)]}{\cosh \kappa d}$$

$$\frac{g\kappa}{\omega^2} a_0 \frac{\sinh[\kappa(\bar{z}+d)]}{\cosh \kappa d} \qquad (3.58)$$

Note that the vertical semi-axis disappears at the bottom. For deep water κd is large, and these expressions become much simplified:

$$\cosh(\kappa d) \simeq \tfrac{1}{2}\exp(\kappa d)$$

$$\cosh[\kappa(z+d)] \simeq \tfrac{1}{2}\exp[\kappa(z+d)] \tag{3.59}$$

$$\sinh[\kappa(z+d)] \simeq \tfrac{1}{2}\exp[\kappa(z+d)] \tag{3.60}$$

The particle motion then decays as $\exp(\kappa z)$ (note that z is negative), and the expression for the velocity potential becomes:

$$\Phi(x, z, t) = i\frac{g}{\omega} a_0 \exp[-i(\kappa x - \omega t)]\exp(\kappa z) \tag{3.61}$$

The dispersion relation becomes:

$$\omega^2 = g\kappa \tag{3.62}$$

$\tanh \kappa d$ tends to 1 for large κd, hence for deep water:

$$\tanh \kappa d \simeq 1 \tag{3.63}$$

Now for $\tanh(2.65) = 0.99$ we require:

$$\kappa d = 2\pi d/\lambda > 2.65 \tag{3.64}$$

So the water may be considered deep if the depth is more than half the wavelength.

One consequence of this is that if $d \gg \lambda/2$ we may neglect the effect of (smooth) variations of the depth d with the horizontal coordinates, provided that $\kappa d > 2.65$ or $d \gg \lambda/2$. We can rewrite (3.62) in terms of the phase velocity c of the waves as either

$$c^2 = g\lambda/2\pi$$

or $$\tag{3.65}$$

$$c = g/\omega$$

We see that the wave phase speed increases with increasing wavelength and decreases with increasing angular frequency ω. As expected from equation (3.62), the higher-frequency waves have shorter wavelengths (Figure 3.6).

Let us now compute the energy contained in one wavelength λ, by considering the potential and kinetic energies of the particles contained within the region $x = 0$ to $x = \lambda$ for unit width of wave crest. The energy can be written $E(\omega, \eta, \lambda)$, where E_λ = kinetic energy

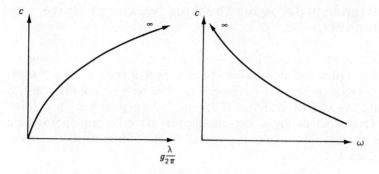

Figure 3.6 *Variation of celerity with wavelength and frequency*

of orbital motion of particles + potential energy due to water-level change. We have:

$$\Phi(x, z, t) = i \frac{g}{\omega} \eta(x) \frac{\cosh\left[\kappa(z+d)\right]}{\cosh \kappa d} \exp(i\omega t) \qquad (3.66)$$

Hence (taking into consideration the kinematic condition and Green's theorem) we can write:

$$E_\lambda = \frac{\rho}{2} \int_0^\lambda \phi \frac{\partial \phi}{\partial n} \, dS + \frac{\rho}{2\kappa} \int_0^\lambda \left(\frac{\partial \eta}{\partial t}\right)^2 dS \qquad (3.67)$$

where $\partial/\partial n$ represents differentiation along the normal to the surface $z = \eta$; E_λ is proportional to the square of the wave 'amplitude' $(\eta_{max})^2$.

The average energy per unit area of the surface area of the wave field, $\bar{E}$, is then given by:

$$\bar{E} = E_\lambda/\lambda$$

For the Airy wave we can easily deduce the average energy below unit horizontal surface area $\bar{E}$ as simply:

$$\bar{E} = PE + KE$$

$$= \tfrac{1}{4}\rho g a_0^2 + \tfrac{1}{4}\rho g a_0^2$$

$$= \tfrac{1}{2}\rho g a_0^2 \qquad (3.68)$$

Invoking the linear approximation, this may now be interpreted as the energy below unit area of the surface itself.

Higher-order wave theories (waves of finite height)

The most commonly used theories for non-linear waves are Stokes's fifth-order and third-order theories. Stokes tried to find a time-harmonic solution to the governing equation $\nabla^2 \Phi = 0$ that satisfied the free-surface conditions (3.18) and (3.24) in their non-linear form. This resulted, for a two-dimensional wave, in an approximate expansion for the velocity potential for the fifth-order theory, of the form:

$$\Phi(x, z, t) = \sum_{\alpha = 1}^{5} \lambda_\alpha \cosh(\alpha \kappa z) \exp[-i\alpha(\kappa x - \omega t)] \qquad (3.69)$$

where the λ_α are constants for a given depth, wave height and wave length. Computation of these constants, however, involves the simultaneous solution of a quintic and a quartic equation, both deduced from the free-surface boundary condition. The resulting motion is irrotational, the particle motions being circular with a superimposed drift velocity. There is a theoretical maximum steepness δ associated with the theory, defined by:

$$\delta = a_o/\lambda < 0.142 \qquad (3.70)$$

Stokes's third-order theory involves an expansion for the velocity potential similar to (3.69), but with only three terms in the expansion. See Skjelbreia et al.[5] for more details of this theory.

Another popular theory is Gerstner's trochoidal wave. This is an exact solution for deep-water waves with finite amplitude. The solution is obtained by changing the frame of reference so that it moves with the waves; the resulting transformed governing equation is then solved with the non-linear forms of the boundary conditions. The resulting motion is rotational, however, which means that the velocity potential cannot be used to represent this type of wave. This introduces severe difficulties into the analysis of any wave–structure interaction problem.

An excellent treatment of the above theories as well as many others can be found in the literature, notably Lamb[1], Stokes[2], Kinsman[3] and Wehausen and Laitone[4].

Comparison of the wave theories

In this book from now on we shall use the linear or Airy theory to represent wave motion, because with this approach there is no interaction between waves of different frequencies, and hence the simple superposition of many different component wave trains of differing frequency and direction is possible without complications. This is not possible with the non-linear wave theories, as the 'mode shapes' corresponding to these theories are not orthogonal. For completeness we here discuss the relative merits of the theories, apart from the above-mentioned overriding consideration.

Dean[6] has compared the various wave theories by calculating the fits to the two free-surface boundary conditions as indicators of the relative validities of the theories. Whereas Laitone[7] used as his criterion a comparison of the wave celerity calculated by the differing theories, Dean himself commented in his paper that 'the errors in quantities of engineering significance associated with errors in boundary condition fits, however, remain to be established'.

Dean[6] calculated the mean square error in the boundary condition equations—the left-hand side minus the right-hand side—over one wavelength, for various heights of waves compared with the breaking wave heights for the different theories. One interesting point, discovered by Dean and mentioned in his paper, is that the Stokes fifth-order non-linear equations which have to be solved to obtain the parameters λ_α have no solution for:

$$h/T^2 < 0.1$$

where T is the time period of the wave; $T = 2\pi/\omega$. For water depth of 100 metres this implies that there are no Stokes fifth-order waves for $\omega < 0.2$. We here reproduce the main results of Dean's paper for reference. The two independent parameters d/T^2 and H/T^2, where H is the trough-to-crest height of the waves, are plotted against each other (Figure 3.7). The lower parts of the bands represent waves far from breaking height, i.e. $H/H_B = \frac{1}{4}$, where H_B is breaking height. The upper parts represent the breaking limit.

As we are interested here in wave motion from the point of view of the wave forces on offshore structures, let us consider the relative magnitudes of these forces with particular reference to the Stokes fifth-order and the linear theories.

In their paper Hogben and Standing[8] made comparisons between

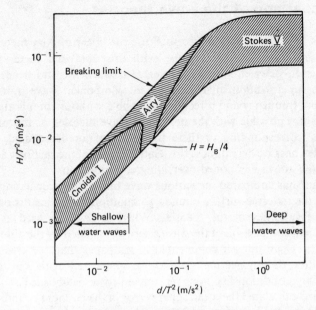

Figure 3.7 Periodic wave theories providing best fit to dynamic free-surface boundary condition (from Dean[6])

linear and Stokes fifth-order waves of the same height H and period T. Values of the parameters H/gT^2 and H/d were chosen to be characteristic of the northern North Sea under extreme weather conditions, as well as for a less steep wave. They chose:

$$H/gT^2 = 0.015 \quad \text{and} \quad 0.01$$
$$H/d = 0.2$$

For the steeper wave the Stokes fifth-order theory gives a wavelength 7 per cent longer than that given by the linear theory. This in itself may be an important difference. Velocity and acceleration profiles were calculated for the two theories, and the results for their maxima are reproduced in Figure 3.8.

It is indicated in this paper that the main effect of non-linearity is to steepen the waves, the differences becoming less important for deeper water. Figure 3.9 shows that the total inertia forces on a slender cylinder calculated by the two theories differ by about 7 per cent whereas the drag contributions differ by less than 13 per cent. Therefore for gravity structures, where inertia forces dominate, the

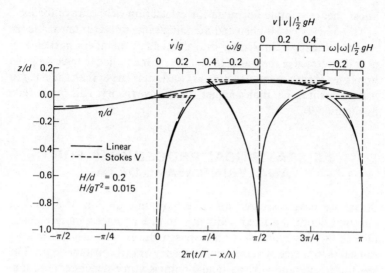

Figure 3.8 Linear and Stokes fifth-order profiles of wave elevation, maximum acceleration and velocity components (from Hogben and Standing[8])

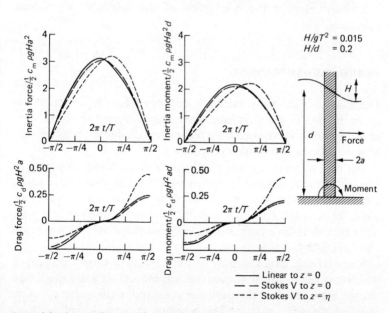

Figure 3.9 Drag and inertia effects (force and overturning moment) on a slender column, according to the linear and Stokes fifth-order theories (from Hogben and Standing[8])

linear theory is quite adequate for calculation of the wave forces.

The conclusion Hogben and Standing came to is that linear theory is adequate except for slender 'drag dominated' members, particularly near the surface of the water. In fact the error entailed in regarding the immersed length of the structure as constant is larger than the error in using a linear rather than a higher-order wave theory. This can be seen in Figure 3.9.

3.3 STATISTICAL PROCEDURES FOR ANALYSING WAVE DATA

So far we have discussed only the harmonic motion of a fluid of constant depth at a fixed frequency. In order to get a more realistic idea of the wave forces on offshore structures, we must not restrict ourselves to waves of a constant frequency or to a constant depth. The sea-surface elevation at a particular point is a time-dependent random variable. We shall now discuss the methods of measurement of this variable and describe some of the statistical measures of its variation commonly used in the calculation of the response of offshore structures.

Wave measurement

There have been many methods developed for the experimental determination of the significant wave height H_s and the mean zero crossing period T_0, the two main parameters needed in the estimation of wave spectra at a particular location.

The first information was obtained by direct visual observation. Paape[10] made a comparison of such observations with a wave gauge and found that H_s was overestimated by this method, typically by about 25 per cent. Later it was found that visual observations were more reliable if some calibrated pole was used against which to judge wave heights.

More reliable data can now be obtained by a variety of gauges, which may use electrical methods such as inductance and capacitance change between two fixed cables or resistance change due to water shorting discrete contacts. The resulting current or voltage fluctuations may be linked to a computer to obtain a wave spectrum directly by the Fast Fourier Transform technique, or to a strip chart for later analysis; 99 per cent accuracy can be obtained by this method.

Another method is to use a floating buoy, which supposedly follows the water surface movement and has within it a number of accelerometers that record its motion. The method is thought to be accurate to about ± 5 per cent.

It is important to realise at this stage that the frequency of the surface elevation fluctuations at a point are independent of the direction of wave propagation past the point of observation, so no measure of the directivity of the wave field can be obtained by any of the methods mentioned so far.

The one remaining popular direct-sensing method is the submerged pressure-sensing method. A simple strain gauge is mounted on an underwater tripod or piling and the pressure at this subsurface point is then measured. This is then correlated to the wave height using linear wave theory and the wave elevations are thus determined. Unfortunately, the slight viscosity and compressibility of the overlying water cause the high-frequency components of the spectrum to be filtered out before reaching the sensing instrument.

Shipborne wave recorders are used, but are inaccurate because the ship itself responds only to the lower-frequency oscillations of the water surface and hence acts as a filter, eliminating part of the available spectral information.

More sophisticated and often laborious techniques are needed when information on the prevailing *direction* of incidence of wave energy is required. For this type of measurement either an array of interconnected sensors is required or more indirect techniques (such as stereophotography, radio-backscatter, radar sensing and even holographic techniques) are used. Holography seems particularly promising, since only aerial photographs are required for this technique. These methods and many more are discussed in Waves 74[11] as well as Kinsman[3].

The probability density function for the wave elevation

Here we consider wind-generated waves only. The so called 'tidal' waves generated by distant earthquakes, eruptions, etc. may be treated deterministically because their occurrence is exceptional, and tides themselves are of such long wavelength as to be considered as (time-independent) currents.

Wind waves are usually generated by storms of some kind, whose dimensions are large in comparison with the wavelengths of the waves considered, and it is admissible therefore to assume that the contributions from different parts of the storm area are of random phase. The range of exciting frequencies in the storm, however, need not be, and usually is not, narrow. The nature of the excitation suggests that we model the random variable (the surface elevation) by an infinite sum of harmonic waves of random phase.

$$\eta(\vec{x}, t) = \sum_{i=1}^{\infty} \eta_i(\vec{x}, t)$$

$$= \sum_{i=1}^{\infty} a_i(\omega_i, \vec{\kappa}_i) \cos(\omega_i t - \vec{\kappa}_i . \vec{x} + \varepsilon_i) \tag{3.71}$$

where $\vec{x} = (x, y)$ is a position vector of the point under consideration

$\vec{\kappa}_i$ is the wavenumber vector of that particular component wave, distributed randomly within the storm area

ω_i are the frequencies of the component waves

ε_i are the random phases distributed *uniformly* on the interval 0 to 2π (i.e. the value of its probability density function is constant and equal to $1/2\pi$)

a_i is the amplitude of each component wave

The above representation of the random variable η, in which the components are considered statistically independent, is exactly the one to which the central limit theorem described in Rice[9] is applicable. We therefore assume that the distribution (spatial and in time) of η is Gaussian. Thus we expect the probability density function of η to be given by:

$$p(\eta) = \frac{1}{\sqrt{(2\pi)M_0}} \exp\left(-\frac{\eta^2}{2M_0^2}\right) \tag{3.72}$$

where $M_0 = \langle \eta^2 \rangle^{\frac{1}{2}}$, the root mean square value of η, and we have assumed $\langle \eta \rangle = 0$, i.e. that surface elevation is measured from the still water level.

Observations of the sea have generally confirmed (3.72), although the assumption breaks down when we consider the fine structure or ripples on the surface. This is not surprising: ripples do not have gravity as the restoring force, surface tension being the dominant effect. Fortunately we are not interested in the effects that ripples have on offshore structures, so we may assume (3.72) to be applicable.

Energy spectra

Considering the amplitudes a_i in (3.71), and recalling that the expression for the energy content per unit area of the sea surface for a harmonic wave gives $E \propto$ (amplitude)^2, we can compute the energy of the waves in the small frequency band $(\omega_i, \omega_i + d\omega_i)$ per unit area of surface, i.e.

$$E(\omega_i, \vec{\kappa}_i) = \tfrac{1}{2}\rho g \sum_{\vec{\kappa}_i}^{\vec{\kappa}_i + d\vec{\kappa}_i} \sum_{\omega_i}^{\omega_i + d\omega_i} a_i^2(\omega_i, \vec{\kappa}_i)$$

$$= \rho g\, S_{\eta\eta}(\omega_i, \vec{\kappa}_i)\, d\omega d\vec{\kappa}_i \qquad (3.73)$$

where $S_{\eta\eta}$ is a continuous function of frequency and direction of wave advance, and is the spectral density function of the random process η.

If we assume an essentially uniform unidirectional wave distribution, as would be produced by a uniform wind (blowing for an infinite time over an infinite ocean and in the positive x direction), we would have:

$$\eta(t) = \sum a_i \cos(\kappa_i x - \omega_i t - \varepsilon_i) \qquad (\omega_i > 0) \qquad (3.74)$$

and

$$E(\omega_i)\, d\omega = \tfrac{1}{2}\rho g \sum_{\omega_i}^{\omega_i + d\omega_i} a_i^2 = \rho g\, S_{\eta\eta}(\omega_i)\, d\omega \qquad (3.75)$$

where $S_{\eta\eta}$ is now the commonly used unidirectional one-sided spectral density function for a 'fully developed sea'. $S_{\eta\eta}$ is sometimes called the energy or power spectrum.

Wave height, particle velocity, spectral densities

We have already deduced the expression for the velocity potential for linear waves, i.e.

$$\Phi(x, z, t) = \frac{ig}{\omega} a_0 \exp[-i(\kappa x - \omega t)] \frac{\cosh[\kappa(z + d)]}{\cosh \kappa d} \qquad (3.76)$$

Generally we have:

$$\eta(x) = -i\,\frac{\omega\phi(x)}{g} = a_0 \exp(-i\kappa x) \tag{3.77}$$

for a unidirectional Airy wave, and:

$$\vec{v} = (v_x, v_y, v_z) = \vec{\nabla}\Phi = \left(\frac{\partial\Phi}{\partial x}, \frac{\partial\Phi}{\partial y}, \frac{\partial\Phi}{\partial z}\right) \tag{3.78}$$

can now be expressed as a function of η.

$$\Phi(x, z, t) = \frac{ig}{\omega}\,\eta(x)\,\frac{\cosh[\kappa(z+d)]}{\cosh\kappa d}\exp(i\omega t) \tag{3.79}$$

Differentiating (3.79), we obtain the particle velocity components:

$$v_x = \frac{g}{\omega}\,\kappa\,\frac{\cosh[\kappa(z+d)]}{\cosh\kappa d}\exp(i\omega t)(x)$$

$$v_y = 0 \tag{3.80}$$

$$v_z = i\,\frac{g}{\omega}\,\kappa\,\frac{\sinh[\kappa(z+d)]}{\cosh\kappa d}\exp(i\omega t)\eta(x)$$

Differentiating again with respect to time, we obtain the components of particle accelerations:

$$\dot{\vec{v}} = (\dot{v}_x, \dot{v}_y, \dot{v}_z) \tag{3.81}$$

which are:

$$\dot{v}_x = ig\kappa\,\frac{\cosh[\kappa(z+d)]}{\cosh\kappa d}\exp(i\omega t)$$

$$\dot{v}_y = 0 \tag{3.82}$$

$$\dot{v}_z = -g\kappa\,\frac{\sinh[\kappa(z+d)]}{\cosh\kappa d}\exp(i\omega t)$$

In Chapter 2 we defined the spectral density of a random process u as:

$$S_{uu}(\omega) = \lim_{T\to\infty}\frac{1}{T}\{\bar{U}\hat{U}\} = \lim_{T\to\infty}\frac{1}{T}|\bar{U}|^2 \tag{3.83}$$

where $\bar{U}$ is the Fourier transform of u and $\hat{U}$ is its conjugate. Applying this to the velocity variables defined in equation (3.80), we obtain:

$$S_{v_x v_x} = \left[\frac{g\kappa}{\omega} \frac{\cosh\left[\kappa(z+d)\right]}{\cosh\kappa d} \right]^2 S_{\eta\eta}$$

$$S_{v_z v_z} = \left[\frac{g\kappa}{\omega} \frac{\sinh\left[\kappa(z+d)\right]}{\cosh\kappa d} \right]^2 S_{\eta\eta}$$

(3.84)

and the corresponding expressions for the accelerations—equations (3.82)—are:

$$S_{\dot{v}_x \dot{v}_x} = \left[g\kappa \frac{\cosh\left[\kappa(z+d)\right]}{\cosh\kappa d} \right]^2 S_{\eta\eta}$$

$$S_{\dot{v}_y \dot{v}_y} = \left[g\kappa \frac{\sinh\left[\kappa(z+d)\right]}{\cosh\kappa d} \right]^2 S_{\eta\eta}$$

(3.85)

Note that the last two spectral densities could be obtained as a special case of formula (2.70), which can now be written:

$$S_{\dot{v}\dot{v}}(\omega) = \omega^2 S_{vv}(\omega)$$

(3.86)

and has been proved in Chapter 2.

The cross spectral densities of the accelerations and velocities will all be zero (assuming that by definition the spectral density is real).

The narrow-band approximation to the spectrum

As we have seen, we would expect the process of wave generation by a storm to produce an essentially broad-band spectrum of waves. We saw, however, that surface waves are essentially dispersive. The longer waves travel faster than the shorter, high-frequency, waves and hence would arrive first at an observation point far from a storm (with respect to the storm's dimensions). Hence the observed spectrum of a distant storm would be composed of those waves whose frequency enables them to travel the intervening distance and arrive at the time of observation. Such an observed wave spectrum would be *narrow* banded. We now examine the expected probability distributions of the maxima of the wave elevation η, i.e. the distributions of wave crests with regard to the narrow-band and broad-band assumptions for a random variable.

To be consistent with the notation of Chapter 2, we write u for the wave elevation because the subsequent analysis holds for a general random variable, which is the superposition of an infinite number of harmonic components with random phase.

Statistical parameters for wave record analysis

We have seen that the sea-surface elevation at a particular location is a time-dependent random variable $\eta(t)$, a function of time whose behaviour may be well represented by the associated spectral density function $S_{\eta\eta}(\omega)$. To fit theoretically derived spectra to the wave environment at a particular location, wave records from that location need to be examined and certain statistical parameters representative of the records need to be extracted and used in the theoretical calculation of the relevant spectral-energy density function. Alternatively, the spectral density may be calculated directly from the observations using analogue or digital computers, but this is not always possible.

Let us now consider some relationships for the wave height and the mean zero crossing period for the cases of narrow- and broad-band spectra. The mean value of the wave height can be deduced using formula (2.137), and is:

$$\bar{H} = \langle H \rangle \simeq 2\sqrt{M_0}\left[\sqrt{(2\ln N)} + \frac{0.5772}{\sqrt{(2\ln N)}}\right] \quad (3.87)$$

where N is the number of maxima and the height H is on the average equal to $2\eta_{max}$. The most probable value for the wave height is, from formula (2.138):

$$H_{mode} \simeq 2\sqrt{M_0}[\sqrt{(2\ln N)}] \quad (3.88)$$

Note that these formulae are not valid for $\varepsilon \to 1$ but they do cover all practical applications. The $H_{1/3}$ or significant wave height H_s is the average of the highest third of the wave heights, and its value can be deduced from Figure 2.28 for any spectral width. (Note that the wave amplitude $\eta = \xi\sqrt{M_0}$ of the figure.)

The average time between successive zero up-crossings, i.e. the *mean zero crossing period* (Figure 3.10), is, from formula (2.111):

$$T_0 = 2\pi\sqrt{(M_0/M_2)} \quad (3.89)$$

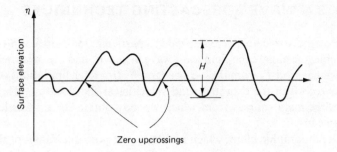

Figure 3.10   Wave height H and zero up-crossings

The average time between successive maxima is, from formula (2.111):

$$T_m = 2\pi \sqrt{(M_2/M_4)} \tag{3.90}$$

Note that the number of maxima can now be calculated, i.e.

$$N = T/T_m \tag{3.91}$$

where T is the duration of the record.

For the narrow-band case, the maxima of the wave elevation follow a Rayleigh distribution, i.e.

$$p(\eta_{max}) = \frac{\eta_{max}}{\sqrt{M_0}} \exp\left(-\eta_{max}^2 \frac{1}{2M_0}\right) \tag{3.92}$$

and by integration it can be shown that:

$$H_{1/3} \simeq H_s = 4\sqrt{M_0}$$
$$\bar{H} = \langle H \rangle = 2.507\sqrt{M_0} \tag{3.93}$$
$$H_{mode} = 2\sqrt{M_0}$$

For the case of a broad-band spectrum, i.e. $\varepsilon = 1$, the distribution of maxima is a Gaussian one, i.e.

$$p(\eta_{max}) = \frac{1}{\sqrt{M_0}\sqrt{(2\pi)}} \exp\left(-\frac{1}{2}\frac{\eta_{max}^2}{M_0}\right) \tag{3.94}$$

3.4 WAVE FORECASTING TECHNIQUES

If an offshore structure is to be sited at a particular location, it is often possible to estimate the significant wave height H_s and the mean zero crossing period T_0 from the local geography (for fetch limitation) and some record of the expected wind velocities at that location. There are three main sources from which we may obtain these statistical parameters:

● Oceanographic maps, which give the most probable values of the height of the highest wave in a 50 or 100 year storm, together with its most probable zero crossing period.

● Wave scatter diagrams, which give information about the different sea states occurring during a certain period, for instance one year. Each sea state is usually defined by its mean zero crossing period (T_0) and the significant wave height (H_s). The number of occurrences is

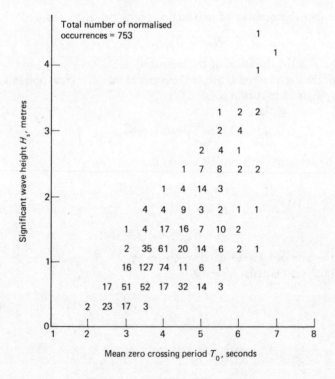

Figure 3.11   Wave scatter diagram

indicated against these two variables to form a diagram such as the one shown in Figure 3.11.

● Wind roses (see Figure 3.12), which give information on the strength, direction and percentage frequency of the expected winds at the location. From these a scatter diagram may be deduced using the curves of Darbyshire and Draper[12], taking into account the local fetch limitations and sea-bed topography if necessary.

This information may be used in a number of ways. In the design wave approach the H_s and T_0 are used directly, whereas if a spectral approach is to be used these parameters define the appropriate

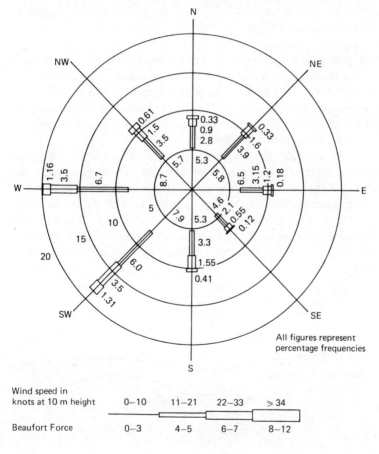

Figure 3.12   Wind rose

spectral form uniquely. For the Pierson–Moskowitz spectrum the only information needed for its definition is the wind speed, so long as the usual empirical constants are used. These 'constants' can, however, be related to H_s and T_0 and thus do in fact vary with the sea state. For some fatigue and foundation settlement problems, a wave-height exceedance diagram may be required to model the extreme conditions of a storm.

The above mentioned forecasting techniques will be discussed in this section. To get an understanding of these problems we shall start by examining the mechanism of wave generation by wind.

Wave generation

The problem of the examination of the mechanism of the transfer of energy from the motion of the air to the sea is a very complex one, and many theories have been formulated as a result. For a full discussion the interested reader is referred to Kinsman[3] and the various journals of advanced fluid mechanics. Some basic features of this process will, however, be discussed briefly here.

Let us consider that a wind of constant mean velocity springs up over an initially calm ocean. The first waves to form will be the ones that have the slowest phase velocity and hence are most able to extract energy from the moving air (Figure 3.13). These (as we saw in section 3.2) are the high-frequency, short-wavelength waves. When formed, they begin to break and in this way transfer energy to lower-frequency components, which are themselves built up by the wind; see Figure 3.14.

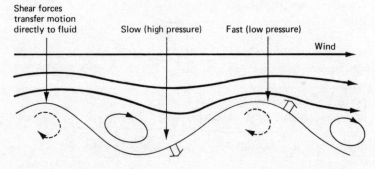

Figure 3.13   Pressure and shear as driving forces on a wind-generated wave

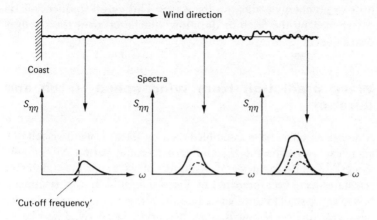

Wind speed	Full spectrum after
10 knots	> 10 nautical miles and > 2.4 hours
20	> 75 and > 10
30	> 280 and > 23
40	> 710 and > 42
50	> 1420 and > 69

Figure 3.14   A developing sea

The short waves are rebuilt, and the energy continues its transfer down the frequency scale until that frequency at which the phase velocity of the wave is equal to the wind velocity. Below this frequency the wind does not supply energy to waves directly, as their phase velocity is higher than the wind speed; most of their energy comes from breaking higher-frequency waves.

It is clear from the above discussion that if the duration of the wind is insufficient for the lower frequency waves to develop, the resulting spectrum will consist of mostly high-frequency components. The sea is then said to be *duration limited*. If there is a coastline upwind of the point of observation, again the low-frequency components will not have time to develop; at this point the sea is said to be *fetch limited*. If the point of observation is far from any coast and the wind has been blowing for a sufficiently long time, we have a fully developed sea.

Pierson *et al.*[13] have shown that, for a wind of a certain duration and fetch, there is a certain cutoff frequency below which there will be little or no energy content in the waves. This cutoff frequency ω_c is determined by the fetch or duration, whichever gives the stronger restriction.

Wave prediction from wind speed, fetch and duration

Many researchers have assembled data on fetch F, wind velocity U, significant wave height H_s, and zero crossing period T_0, notably Wiegel[14] and Bretschneider[15]. A number of wave spectra for different sea states have been prepared by Pierson et al.[13]. In this treatment, however, we shall follow Darbyshire and Draper[12] in determining the design parameters H_s and T_0, since the curves they derived were based on considerably more observations than any of the others.

The graphs presented here (Figures 3.15 and 3.16) give the maximum wave height H_{10} occurring in a typical wave record of ten minutes in length, i.e. the distance from trough to crest of the highest wave of the record. The significant wave height H_s, defined earlier, is related to H_{10} by the equation:

$$H_{10} = 1.60 H_s \tag{3.95}$$

This equation is only strictly correct for a narrow-band record, but is applicable for records with a wider frequency range (Tucker[16]). The curves in Figure 3.15 represent the results obtained for oceanic waters, the curves of Figure 3.16 being for coastal waters 30–45 metres deep. No allowance has been made for refraction, reflection and diffraction of waves.

The shallow-water curves only take into account the change in wavelength and celerity of waves due to the sea bottom, and should not be taken as representative of the waves encountered near shoals, cliffs or bays.

Given the wind speed, fetch and wind duration, the maximum wave height for a ten-minute record may be read from either Figure 3.15 or 3.16. Starting from the surface wind speed in knots (1 knot = 1.15 mile/h = 0.515 m/s) on the vertical axis, follow a horizontal straight line until the duration limit or the appropriate fetch limit position is reached (stopping at whichever is reached first). The maximum wave height H_{10} may then be read off from the curve nearest to the point reached.

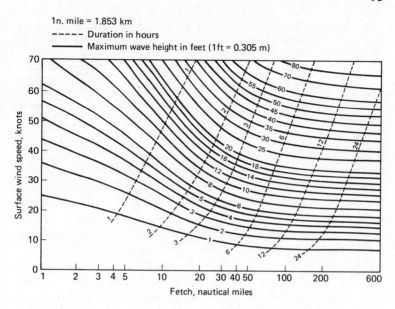

Figure 3.15 H_{10} *for oceanic waters (from Darbyshire and Draper*[12]*)*

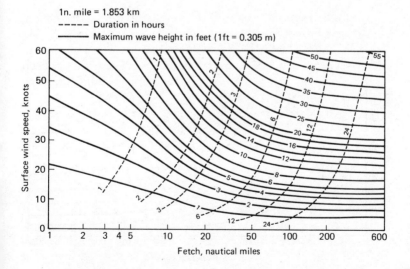

Figure 3.16 H_{10} *for coastal waters (from Darbyshire and Draper*[12]*)*

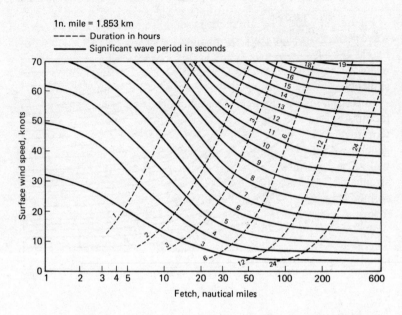

Figure 3.17   Significant wave period for oceanic waters (from Darbyshire and Draper[12])

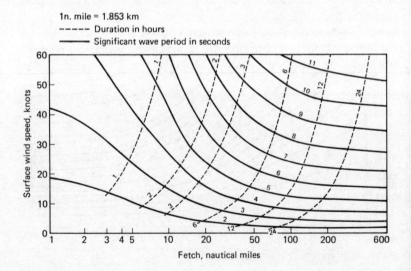

Figure 3.18   Significant wave period for coastal waters (from Darbyshire and Draper[12])

Figures 3.17 and 3.18 give the significant wave period T_s, that is the wave period corresponding to a regular wave with the significant wave height H_s. This value may be taken to be approximately equal to the mean zero crossing period T_0 in most situations. It is obtained in a similar way to H_{10}, using the wind speed, duration and fetch.

In coastal waters the fetch is usually determined by geographical factors, but in the open ocean the fetch limitation may be meteorological, i.e. the wind direction may change over the wave-generating region. In addition, swell waves generated in other regions will be present in the area. For a distant storm the waves will be attenuated by spreading and friction; a good attenuation factor commonly used is $\sqrt{(300/R)}$, where R is the distance in nautical miles from the storm centre.

When a swell of height h_{swell} is to be added to the calculated wave height due to the local winds h_{local}, the total wave height H_{tot} is given by:

$$H_{tot} = \sqrt{(h_{swell}^2 + h_{local}^2)} \qquad (3.96)$$

h_{local} is calculated by the method described above.

In some circumstances, such as in fatigue calculations, the designer may require some idea of the magnitude of the highest wave in a storm. Clearly, the longer the storm lasts the greater the chance of constructive interference between the various wave components, and hence the greater will become the highest wave encountered during the storm. The factor by which H_{10} is to be multiplied is given in Figure 3.19 as a function of storm duration; these curves are deduced from expressions derived in Longuet–Higgins[17]. This relationship between the significant wave height H_s and the maximum H_{max} in a record of N cycles is given by:

$$H_{max} = H_s\sqrt{(0.5 \ln N)} \qquad (3.97)$$

This relationship is based on the Rayleigh distribution and hence assumes a narrow-band process. As the square of the wave periods T also follows a Rayleigh distribution, we may write:

$$T_{max}^2 = T_s^2\sqrt{(0.5 \ln N)} \qquad (3.98)$$

where T_{max} is the maximum wave period in a record of N cycles and T_s is the significant period derived earlier.

If we have a narrow-band process for which we may assume a Rayleigh distribution for T^2 and H, it is possible to calculate the

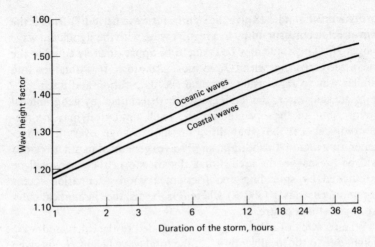

Figure 3.19 *Wave-height factor for storms of various durations (from Darbyshire and Draper[12])*

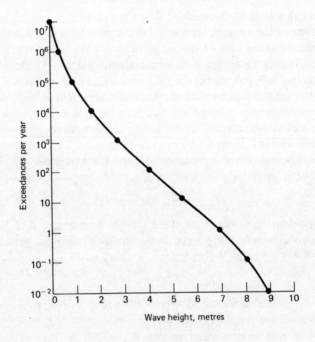

Figure 3.20 *Height exceedance diagram*

probabilities of exceedance of a particular wave height and period. For a Rayleigh distribution:

$$P(H) = \frac{H}{M_0} \exp\left(-\frac{H^2}{2M_0}\right) \tag{3.99}$$

This gives, for the probability that $H > \overline{H}$:

$$P(H > \overline{H}) = \int_{\overline{H}}^{\infty} \frac{H}{M_0} \exp\left(-\frac{H^2}{2M_0}\right) dH$$

$$= \begin{cases} 1 & \overline{H} \leqslant 0 \\ \exp\left(-\frac{H^2}{2M_0}\right) & \overline{H} \geqslant 0 \end{cases} \tag{3.100}$$

These probabilities may be converted into numbers of waves by multiplying by the record length to give the time for which, say, $H > \overline{H}$, and dividing by the zero crossing period. If these wave exceedances are plotted on log/linear graph paper the resulting curve is usually a straight line, which may be easily extrapolated. The usual record length is one year. An example of a height exceedance diagram is given in Figure 3.20.

Derivation of a scatter diagram from a wind rose

In this section we describe how a scatter diagram (such as the one in Figure 3.11) for a particular location may be deduced from a wind rose (Figure 3.12). For the purposes of the construction of a scatter diagram, we ignore the direction of the wind given by the wind rose and neglect any effects that result from the interference of waves travelling in different directions over the same body of water. This is admissible in the linear approximation applied here.

The wind rose shown in Figure 3.12 will give a total of 32 points on our scatter diagram, one for each wind strength in each direction. The percentage frequencies associated with each wind strength in a given direction may be reinterpreted as occurrences out of 100 of this wind speed. For each point we use the Darbyshire and Draper curves to determine H_s and T_s (or T_0) corresponding to this wind speed, and enter the number of occurrences at this point on the scatter diagram at the appropriate position. Using Figure 3.12 would give us a total number of occurrences of 100. If two points so obtained lie close together on the scatter diagram they may simply be added and repositioned at some mean position.

Scatter diagrams are also obtained directly from wave measurements taken usually over a period of more than one year. Probabilities of wave-height exceedance may be taken from this scatter diagram directly, and the height exceedance diagram may be constructed as described in the previous section using the zero crossing periods from the same diagram.

Two commonly used spectral formulae

(a) *Pierson–Moskowitz*[19]. Once H_s and T_0 (the mean wave period) have been estimated we may use the following empirically derived expression for the spectral energy density function $S_{\eta\eta}(\omega)$, derived by Pierson–Moskowitz as:

$$S_{\eta\eta}(\omega) = \frac{\alpha g^2}{\omega^5} \exp\left[-\beta\left(\frac{g}{\omega W}\right)^4 \right] \qquad (3.101)$$

where W is the wind speed for which the spectrum is required
  g is the acceleration due to gravity
  α and β are dimensionless constants dependent on H_s and T_0
  and given by the following equations:

$$\alpha = 4\pi^3 \left(\frac{H_s}{g T_0^2}\right)^2 \qquad \beta = 16\pi^3 \left(\frac{W}{g T_0}\right)^4 \qquad (3.102)$$

or alternatively:

$$H_s = \frac{2W^2}{g}\frac{\alpha}{\beta}$$
$$T_0 = 2\pi \frac{W}{g} \frac{1}{(\beta\pi)^{\frac{1}{4}}} \qquad (3.103)$$

We also have the moments $M_0 = W^4\alpha/4\beta$ and $M_2 = \alpha W \sqrt{\pi}/4$. It can be seen that α and β depend on the wind speed W (as do H_s and T_0), but the dependence is weak. For the North Sea α and β are taken as 0.0081 and 0.74 respectively. The wind speed W is conventionally taken at 19.5 metres above the still water level.

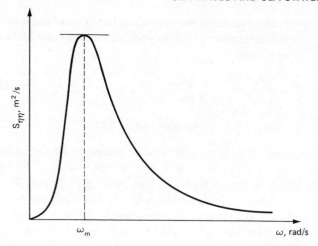

Figure 3.21 Pierson–Moskowitz power spectral wave-height density

As can be seen from Figure 3.21 the Pierson–Moskowitz spectrum is applicable only for a fully developed sea, since the lower frequencies are present. A full account of the derivation and properties of this spectrum may be found in Pierson and Moskowitz[18].

(b) *JONSWAP (Joint North Sea Wave Project).* This spectrum was postulated in an attempt to take into account the higher peaks of spectra in a storm situation for the same total energy as compared with P–M, and also the occurrence of frequency shifts of the maximum in these circumstances. This spectrum is still unimodal and is supposed to represent the wave conditions in a fetch-limited sea; the incorporation of the overshoot parameter γ does go some way towards modelling the worst-case spectrum encountered during a typical storm. The functional form of this spectrum is given by:

$$S_{\eta\eta}(\omega) = \frac{\alpha g^2}{\omega^5} \exp\left[-\frac{5}{4}\left(\frac{\omega_m}{\omega}\right)^4 \right] \gamma^{\exp[-(\omega - \omega_m)^2/2\sigma^2\omega_m^2]} \quad (3.104)$$

where α is the same parameter used in the P–M spectrum

g is the acceleration due to gravity

ω_m is the frequency corresponding to maximum energy density given by the P–M spectrum, i.e.

$$\omega_m = \left(\frac{4}{5}\beta\right)^{\frac{1}{4}} \frac{g}{W}$$

γ is the overshoot parameter, i.e. the ratio of the maximum spectral energy to the corresponding maximum of the P–M spectrum

$$\sigma = \begin{cases} \sigma_a & \text{for } \omega < \omega_m \\ \sigma_b & \text{for } \omega > \omega_m \end{cases} \tag{3.105}$$

σ_a = left side width
σ_b = right side width (see Figure 3.22)

The average values of σ_a and σ_b were found to be 0.07 and 0.09 respectively. Typically $1 < \gamma < 7$, with a mean value of 3.3 for the North Sea. γ is found from T_p (the peak period) and H_s using Figure 3.23, from Chakrabarti and Snider[19]. A comparison of the two spectral shapes in Figure 3.24 shows clearly the higher peak of the JONSWAP spectrum and its shift towards higher frequencies; this gives rise to the so-called overshoot, indicated in the figure, which is inherent in the P–M (fully-developed sea spectrum). This effect is

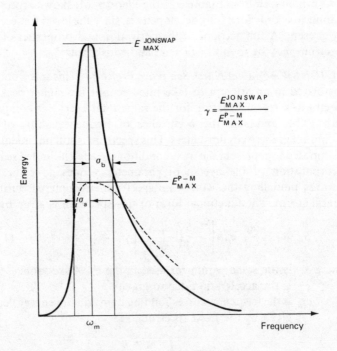

Figure 3.22   The JONSWAP spectrum

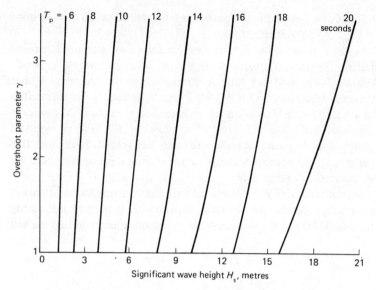

Figure 3.23 Significant wave height as a function of overshoot parameter and peak period, for the JONSWAP spectrum (from Chakrabarti and Snider[19])

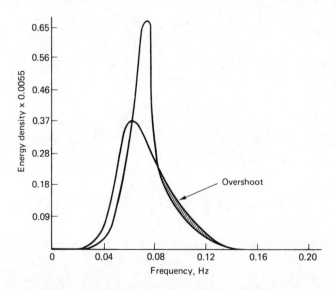

Figure 3.24 Shift in peak frequency between P–M and JONSWAP ($\gamma = 3.3$) spectra for the same significant wave height (from Chakrabarti and Snider[19])

thought of as being due to non-linear interactions taken into account by the JONSWAP spectrum.

In their paper Chakrabarti and Snider have presented various spectra observed at various stages of development of a North Atlantic storm in March 1968. A comparison of the measured H_s and predicted H_s by both the JONSWAP and P–M models is presented in Figure 3.25, and it clearly shows the lag inherent in the P–M model as compared with the more closely following JONSWAP spectral model. Three actual spectra are also compared with the two theoretical ones at different stages in the storm's development, and are reproduced in Figure 3.26.

Obviously JONSWAP keeps up with the evolving storm situation much more closely than Pierson–Moskowitz. It may be fair to say that the JONSWAP spectrum does not model a fetch-limited sea but

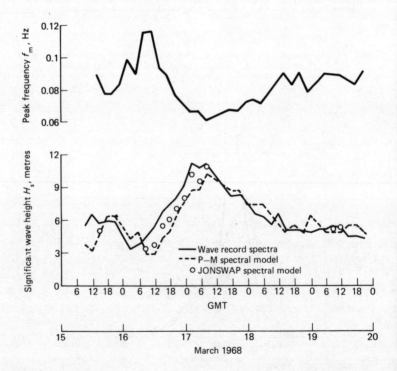

Figure 3.25   Comparison of measured spectral H_s and JONSWAP and P–M H_s versus time for 1968 North Atlantic storm (from Chakrabarti and Snider[19])

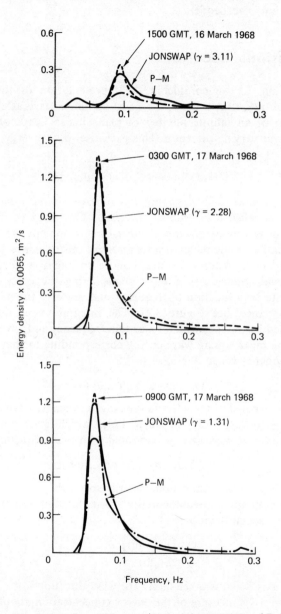

Figure 3.26   Sample plots, from 1968 storm, of recorded JONSWAP and P–M spectra

more exactly a duration-limited one, although in practice often the distinction is academic.

Directional spectra

In section 3.3 we considered the representation of the surface elevation $\eta(\vec{x}, t)$ at a point $\vec{x} = (x, y)$ on the surface as the superposition of an infinite number of plane linear waves of random phase and varying direction. This gave rise to an energy spectrum $S_{\eta\eta}(\omega_i, \vec{\kappa}_i)$, defined by:

$$E(\omega_i, \vec{\kappa}_i) = \rho g S_{\eta\eta}(\omega_i, \vec{\kappa}_i) \, d\omega_i \, d\vec{\kappa}_i$$

The spectrum so defined is a function of frequency *and wave direction* $\vec{\kappa}_i$, and is called a directional spectrum. This form of directional spectrum is, however, rather cumbersome to manipulate, since the wavenumbers $\vec{\kappa}_i$ are themselves functions of the frequency ω_i through the dispersion relation. A much more manageable form of the directional spectrum is given by considering the energy spectral density to be a function of frequency and angle to the direction of wave advance. See Figure 3.27. The integral under the surface generated by the spectrum for a certain frequency and angular range gives the mean square wave energy corresponding to that direction and frequency range. We have:

$$E(\omega_i, \theta) = \rho g S_{\eta\eta}(\omega_i, \theta) \, \omega \, d\omega \, d\theta \qquad (3.106)$$

If we set ω equal to some fixed value, say ω_0, then we have a function $S_{\eta\eta}(\omega_0, \theta) = S_{\eta\eta}(\theta)$, say, of angle only, which gives the angular distribution of wave energy associated with that frequency. Then:

$$S_{\eta\eta}(\omega, \theta) = S_{\eta\eta}(\theta) S_{\eta\eta}(\omega) \qquad (3.107)$$

where $S_{\eta\eta}(\omega)$ is a unidirectional energy spectrum.

One commonly used theoretical form for this function is the circular normal function:

$$S_{\eta\eta}(\theta) = \frac{\exp[a\cos(\theta)]}{2\pi I_0(a)} \qquad \mathrm{m^2\,s^{-1}\,rad^{-1}} \qquad (3.108)$$

where axis $\theta = 0$ is taken along the wind direction

$\quad\quad a$ is a measure of the energy concentration about this axis

$\quad\quad I_0$ is the modified Bessel function of zero order

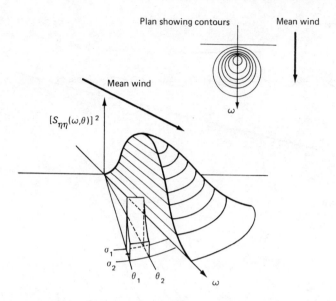

Figure 3.27   Two-dimensional spectrum (from Kinsman[3], courtesy Prentice-Hall Inc.)

Further information on directional spectra can be found in Borgman[20], Kinsman[3] and Panicker[21].

Effects of shallow water

So far we have considered the effects that local geography and varying wind speed have on the expected wave energy spectra at a particular point. We have, however, omitted one aspect of the theory that is often important in offshore engineering, that is the effect that a variation in water depth has on the wave climate at a particular location.

We may be in the position of being able to estimate the wave spectrum at a particular point in deep water and need to estimate how this spectrum evolves as the waves come inshore, perhaps towards a proposed structure we need to analyse.

The effect of variable depth on a regular wave train is well known. As the waves head towards the beach the front face of the wave will steepen and the wavelength shorten until the wave eventually 'breaks' on the shoreline. In addition the wave crests will be refracted until

Side view

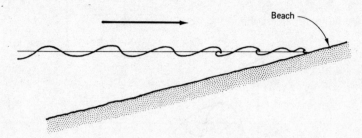

Plan view

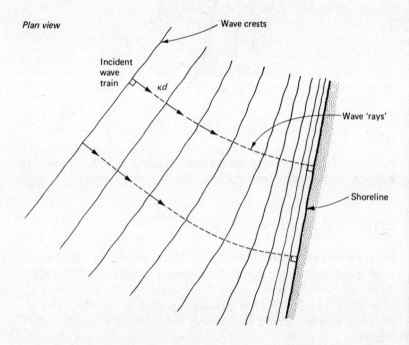

Figure 3.28   Regular waves approaching a sloping shore

they are parallel with the shoreline, so we have a change of direction and wavelength. The frequency and period of the waves are, of course, constant. See Figure 3.28.

The main result derived by Tayfun, Yang and Hsiao[22] of the effect of depth variations is presented below. We consider the effect on a unidirectional spatially homogeneous wave field $S_{\eta\eta}(\omega)$ of shoaling (refraction by the varying depth). This need not be caused by a beach; we may here be considering a submerged feature of the sea bed.

We have, for deep water:

$$\omega^2 = g\kappa_d \tag{3.109}$$

and, for the shallower water of depth $d(\bar{x})$ at $\bar{x}$:

$$\omega^2 = g\kappa_s \tanh\left[\kappa_s d(\vec{x})\right]$$

Hence:
$$\kappa_d = \kappa_s \tanh\left[\kappa_s d(\vec{x})\right]$$

We can now write our spectral density function in deep water, using (3.109) in terms of the wavenumber, as:

$$S_{\eta\eta}^*(\kappa_d) = S_{\eta\eta}(\omega) \tag{3.110}$$

Then it can be shown (Collins[23]) that:

$$\frac{S_{\eta\eta}(\vec{x}, \kappa_s)}{\kappa_s} = \frac{S_{\eta\eta}^*(\kappa_d)}{\kappa_d} \tag{3.111}$$

where $S_{\eta\eta}(\vec{x}, \kappa_s)$ is our spectral density function at depth d, position $\vec{x}$. Hence:

$$S_{\eta\eta}(\bar{x}, \kappa_s) = \coth\left[\kappa_s d(\vec{x})\right] S_{\eta\eta}^*(\kappa_d) \tag{3.112}$$

This gives us a measure of the change in wavelength of the waves along a wave ray.

The unidirectional energy spectrum $S_{\eta\eta}(\omega)$ expressed in terms of the frequency should theoretically stay the same during approach to the shallow water, since each component harmonic function of the surface elevation retains its energy and conserves its frequency for the linear approximation. If wave breaking occurs, however, energy will be redistributed generally from the low-frequency long waves to the higher-frequency waves, giving a shift of energy upwards in the range.

For unidirectional spectra it may be necessary to change the implied wave direction by wave tracing, as indicated in Figure 3.28. A representative wave frequency must be chosen for this process, however, because waves of different lengths are refracted differently. For a fuller treatment see Kinsman[3].

References

1. Lamb, H., *Hydrodynamics*, 6th edn, Cambridge U.P. (1962)
2. Stoker, J. J., *Water waves*, Interscience Publishers (1957)
3. Kinsman, B., *Wind waves, their generation and propagation on the ocean surface*, Prentice-Hall (1965)
4. Wehausen, J. V., and Laitone, E. V., 'Surface waves', *Encyclopaedia of physics*, Vol. 9, Springer, Berlin (1960)
5. Skjelbreia, L., and Hendrickson, J., 'Fifth-order gravity wave theory', Proc. Seventh Conf. on Coastal Engineering (1961)
6. Dean, R. G., 'Relative validities of water wave theories', *Proc. ASCE (Waterways and Harbors Div.)*, **96**, WW1 (Feb. 1970)
7. Laitone, E. V., 'Limiting conditions for cnoidal and solitary waves', *J. Fluid Mechanics*, **9**, Part 3, 1555–1564 (Nov. 1960)
8. Hogben, N. and Standing, R. G., 'Experience in computing wave loads on large bodies', OTC 2189, Proc. Offshore Technology Conf., Houston (1975)
9. Rice, S. O., 'The mathematical analysis of random noise', in *Selected papers on noise and stochastic processes*, ed. Wax, N., Dover, New York (1954)
10. Paape, A., 'Some aspects of the design procedure of maritime structures', 22nd Int. Navigation Congress, Paris (1969)
11. *Proc. Int. Symp. on Ocean Wave Measurement and Analysis* (2 volumes), ASCE (1974)
12. Darbyshire, M., and Draper, L., 'Forecasting wind-generated sea-waves', *Engineering* (London), **195**, 5 (Apr. 1963)
13. Pierson, W. J., Neumann, G., and James, R. W., *On ocean wave spectra and a new method of forecasting wind-generated sea*, US Army Corps of Engineers, Beach Erosion Board, Tech. Mem. No. 43 (1953)
14. Wiegel, R. L., *Oceanographical engineering*, Prentice-Hall (1964)
15. Bretschneider, C. L., 'The generation and decay of wind waves in deep water', *Trans. Am. Geophys. Union*, **33**, No. 3, 381–389 (June 1952)
16. Tucker, M. J., 'The measurement of the height and period of sea waves', *Proc. Instn Civil Engineers* (1963)
17. Longuet-Higgins, M. S., 'On the statistical distribution of the heights of sea waves', *J. Mar. Res.*, **11**, 245–266 (1952)
18. Pierson, W. J., and Moskowitz, L. 'A proposed spectral form for fully developed wind seas based on the similarity theory of S. A. Kitaigorodskii', *J. Geophys. Res.*, **69**, 5181 (1964)
19. Chakrabarti, S. K., and Snider, R. H., 'Modelling of wind waves with JONSWAP spectra', *Proc. Modelling*, 120–139 (1975)
20. Borgman, L. E., *The estimation of parameters in a circular normal two-dimensional wave spectrum*, Tech. report HEL 1–9, Hydraulic Engineering Laboratory, Univ. California, Berkeley (1967)
21. Panicker, N. N., 'Review of techniques for directional spectra', *Proc. Int. Symp. on Ocean Wave Measurement and Analysis*, ASCE, Vol. I (1974)
22. Tayfun, M. A., Yang, C. Y., and Hsiao, G. C., 'Design considerations on wave spectrum estimation', *Proc. Int. Symp. on Ocean Wave Measurement and Analysis*, ASCE, Vol. I (1974)
23. Collins, I. J., 'Prediction of shallow water spectra', *J. Geophys. Res.*, **77**, No. 15 (1969)

4 Forces on Slender Members

4.1 DYNAMIC FORCES

In this chapter we shall examine the various forces that arise from the interaction of the structure with its environment, particularly the fluid–structure interaction.

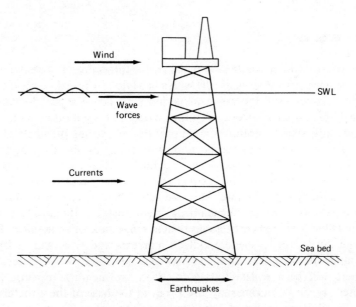

Figure 4.1 Environmental loads on typical offshore structure

Many environmental loads have a time-dependent behaviour whose time scale is comparable with the resonant period of the structure, and hence a full dynamic analysis is required. In addition these loads are random in nature, which necessitates a random vibration analysis because only statistical measures of their variation can be known. (See Chapter 3 and Figure 4.1.)

Wind forces

The environmental forces most familiar to civil engineers are of course the forces due to the wind. Spectra for these forces can be calculated from wind fluctuation spectra, a value for the mean wind speed and its profile. This topic will be dealt with in full in Chapter 6. For the present it is sufficient to note that these forces are only 5–10 per cent of the total environmental forces on a typical offshore structure, and of course act totally on the superstructure, often in the same direction as the predominant wave forces. Because of the low value for the density of air, these forces take the form of drag not inertia terms in our equations.

Currents

Also dealt with in detail in Chapter 6 are the forces due to currents and tides. These processes generally have a behaviour whose time period is much larger than the resonant period, and hence a pseudo-static analysis may seem adequate. Unfortunately this is not often the case, since any 'steady' motion of fluid past the supporting members of a structure gives rise to the phenomenon of vortex shedding by the members; see Figure 4.2. The detachment of a vortex from the member leaves behind it an equal and opposite circulation of fluid round the cylinder; by a well known result of fluid dynamics, this gives rise to a lift force on the member at right angles to the direction of fluid flow. The shed vortices may then impinge on another member. In addition to vortex-shedding effects, currents and tides will in fact change the wave field on the surface and thus alter the wave forces, there will be a constant drag force on the members because of viscosity, and scouring may also occur at the base of the structure.

In addition to a full appreciation of the above concepts, the

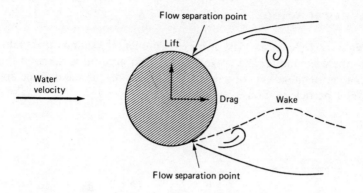

Figure 4.2   Vortex shedding

designer should also be prepared to take into account the effects that marine growth on the structure will have on his calculations, and should also consider the effects of water-temperature variations. Hydrostatic pressure must also be taken into account, since its presence may have the effect of inducing buckling, particularly in the more deeply immersed members.

Earthquakes

Often offshore structures have to be built in seismically active regions in order to exploit the indigenous resources. The response of the structure to earthquakes must then be taken into account.

In the analysis (Penzien *et al.*[1]) the ground acceleration is taken to be a zero mean ergodic process of finite duration. The additional assumption of stationarity of the process may be introduced if the autocorrelation function in time is reasonably small for time separations of 20–50 seconds. If this is not the case, the analysis may be performed with step-by-step integration in the time domain, using an actual record of the earthquake.

If we assume stationarity we may use a spectral approach. Once we have assumed a form for the power spectral density of the ground acceleration $\ddot{u}_g$, we must then take into account the drag, wave radiation damping and inertia effects of the water surrounding the structure. An analysis of this problem, closely following Penzien's, is presented in the last section of this chapter.

Wave loading

Now we come to that part of the environmental loading attributable to the wave motion, i.e. wave loading. (No account is taken in this chapter of the effects of winds and currents on the wave shape and water particle motions.) See Figure 4.3.

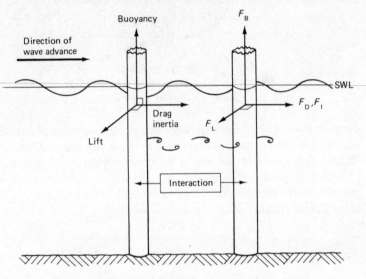

Figure 4.3   Wave forces due to two-dimensional Airy wave

The most obvious result of wave motion past a vertical member is that there is a fluctuating buoyancy force in the vertical direction because of the varying length of the member immersed. This force can be calculated from simple hydrostatics, and is given by:

$$F_B = \frac{b\rho g a_0}{\kappa}\left[\sin(\kappa b - \omega t) + \sin \omega t\right] \qquad (4.1)$$

for a square-sectioned member of width b in the presence of a sinusoidal Airy wave of amplitude a_0, angular frequency ω, and wavenumber κ. The origin is at the front face of the member, and $t = 0$ when the crest has just passed the member and the surface elevation at $x = 0$ is zero. For other shapes, the calculation of this force just involves the evaluation of the immersed volume of the member at any time during the motion.

Calculations of the above sort do not take into account the effects of wave steepness and, for horizontal members near the still water level (SWL), the effects of wave impact or 'slamming'. The solution usually adopted is to avoid the siting of such members in the wave zone. We see from Figure 4.4(a) that the slamming force consists of two parts:

● the fluctuating buoyancy force due to the periodic immersion of the member, and

● the force arising from the transfer of the upward momentum of the water particles in the part of the wave that was below the member.

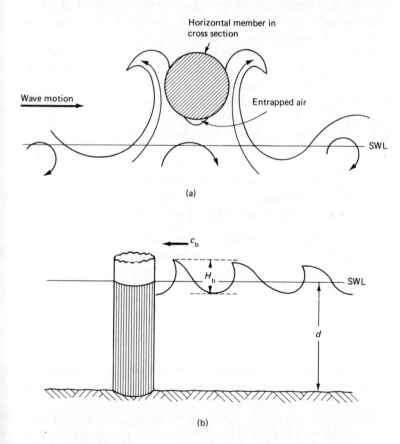

Figure 4.4   (a) Wave slamming on horizontal member; (b) wave slapping on vertical member

There are so many variables involved here that no simple analysis is possible without a great many assumptions about the wave geometry, crest length etc.

Vertical members may experience horizontal impact forces, which are more important the nearer the diameter of the member is to the wavelength considered. The effect of wave steepness can to a certain extent be incorporated into the empirical coefficients (as mentioned by Morison et al.[2]). However, as the wave steepens, in breaking for example, the impact force greatly exceeds the force corresponding to the orbital velocity, because in this condition the water particles above the SWL are largely travelling with the wave velocity. For larger members there may be a cushion of air trapped between the wave front and the member; in the subsequent impact this bubble will 'explode', sending a plume of water up the face of the member.

For the purposes of a rough calculation we may adopt the following assumptions (see Muir Wood[3]):

● At breaking, 75 per cent of the height of the wave is above SWL; see Figure 4.4(b).

● The forward velocity of the water particles is:

$$V_b \simeq c_b = \sqrt{(gd)} \qquad (4.2)$$

where c_b is the velocity of the breaking wave
g is the acceleration due to gravity
d is the undisturbed depth of the water

The force on the face of the member (assuming its dimensions to be comparable to the wavelength) is then equal to the rate of destruction of momentum of the water particles. Hence the (dynamic) pressure on the member is given by:

$$P_b \simeq \rho V_b^2 = \rho dg \qquad (4.3)$$

where ρ is the density of water.

The two main identifiable wave forces on a supporting member of an offshore structure are the drag and inertia forces; these are both in line with the wave direction.

The *drag force* is simply due to the effects of viscosity in the fluid (discounting the vortex shedding effects), and is commonly assumed to be proportional to the square of the water particle velocity relative to the member of the point under consideration.

The *inertia force*, which is independent of any viscosity present, can be thought of as being composed of two parts:

● Hydrodynamic, added, or virtual mass of the member in motion, representing the part of the water that is 'entrained' with the moving member. This added mass has the effect of increasing the apparent mass of the member, and can be thought of as an added inertia or a force in opposition to the motion of the member (see Figure 4.5).

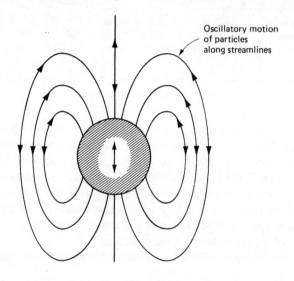

Oscillatory motion of particles along streamlines

Figure 4.5   Streamlines of flow around oscillating cylinder (in frame of reference moving with cylinder)

● The inertia force on a stationary member in an accelerating fluid. This force arises because the fluid has to flow round the member, and hence the streamlines are distorted (shown in Figure 4.6 for a vertical cylindrical member in plan view). The distortion betrays the presence of a pressure gradient within the accelerating fluid, resulting in a force that is communicated to the structure across its solid surface. This force is analogous to the buoyancy force that exists in the vertical direction.

The two (horizontal) forces mentioned above, together with lift forces, will be dealt with quantitatively in section 4.2.

Another important effect of fluid–structure interaction is radiation damping. The oscillating structure excited by waves, earthquakes,

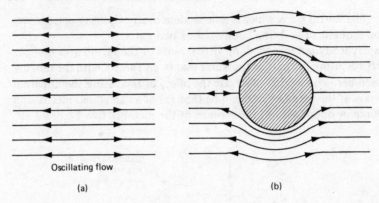

Oscillating flow

(a)   (b)

Figure 4.6   Streamlines of flow (a) in absence of and (b) around a stationary cylinder in accelerating fluid

vortex shedding etc. will itself produce waves on the sea surface; these will themselves carry momentum away from the structure, resulting in a reaction on the motion of the structure. Obviously the larger the typical dimensions of the structure and amplitude of oscillation the larger will be the amplitude of the resulting waves, but because of the dispersion relation (in the Airy wave approximation, see equation (3.51)) the waves will be only of a given wavelength and frequency determined by the motion of the structure.

From the above-mentioned equation, if the structure is oscillating with a given frequency ω we have:

$$\omega^2 = g\kappa \tanh \kappa d \tag{4.4}$$

where κ is the wavenumber $= 2\pi/\lambda$

d is a characteristic depth of the fluid near the structure

For deep water ($\kappa d > 2.6$):

$$\omega^2 \simeq g\kappa \tag{4.5}$$

and we obtain radiated waves of wavelength λ, where:

$$\lambda = 2\pi g/\omega^2 \tag{4.6}$$

For slender members (as defined in the next section) we may ignore this effect, but we return to it at the end of Chapter 5, where diffraction effects are examined in detail for large-diameter members.

4.2 WAVE FORCES ON STATIONARY SLENDER MEMBERS

In this section we look at the various regimes in which various simplifying assumptions can be made. Having made the relevant assumptions we then carry through the analysis in full, in particular for the drag, lift and inertia forces on a stationary cylinder in accelerating fluid flow, using largely the Morison equation (Morison et al.[2]).

For members for which the ratio of significant length (the diameter, in the case of a cylinder) to wavelength is small, one can use a Morison-type formula to obtain the forces. This formula gives inertia and drag forces without considering any modification in the shape of the wave, and is generally accepted to be valid for a diameter-to-wavelength ratio of less than 0.2. Structures of large diameter (such as some of the gravity structures being built) are outside this range, and we then have to take into consideration the change in the shape of the waves due to diffraction effects. For large-diameter structures the drag components are very much less important than the inertia ones, which tend to be large. It is these inertia effects that the diffraction theory, which is valid for inviscid fluids (i.e. no drag), is able to calculate.

Predominance of inertia or drag

Hogben[4] has delineated the various regimes in which the drag and inertia forces described in section 4.1 dominate, with respect to the two parameters H (wave height, trough to crest) and D (typical linear dimension of the intersection of the SWL with a structure, for instance the diameter of a circular cylinder). In addition, in the above-mentioned paper Hogben has tried to delineate the regimes where diffraction-theory methods are necessary in the wave-force analysis. These are reproduced in Figure 4.7.

The upper part of Figure 4.7 indicates the range of dimensions of various structures, related to the horizontal scale of the lower part of the figure. This lower part indicates the ranges of the parameters D and H in which certain forces dominate or in which certain methods of analysis are necessary. The diagonal lines from bottom left to top right, marked 10 per cent drag and 90 per cent drag, refer to the ratio

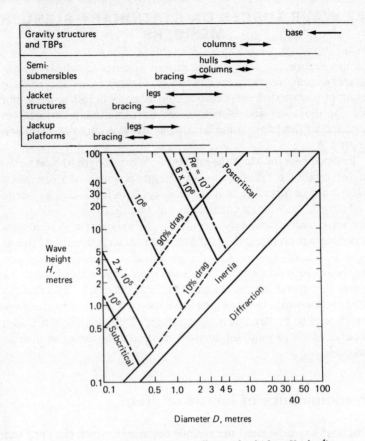

Figure 4.7 Loading regimes at still water level (from Hogben[4])

of the drag to inertia forces along these contours. The diagonal lines from top left to bottom right are contours of Reynolds number, defined by:

$$Re = VD/v \qquad (4.7)$$

where D is a typical length associated with the horizontal dimensions of the structural members at the SWL

V is a characteristic velocity of the flow

v is the kinematic viscosity of the fluid, which is the viscosity of the fluid (usually written μ) divided by its density ρ

Reynolds number is usually used for steady flow, so its significance in this context is somewhat uncertain.

Hogben has here considered a wave with wavelength $\lambda = 15H$, so the usual criterion for determining whether diffraction theory is necessary, i.e.

$$D > 0.2\lambda \tag{4.8}$$

takes the form:

$$D > 3H$$

Figure 4.7 is only meant to be a rough guide, because, even if there is no significant disturbance of the incident field by any particular structural member, to take into account its interaction with neighbouring members we must use the full diffraction theory.

Care should be taken in using the figures when considering a spectral approach, since effectively a typical (design) wave with $\lambda = 15H$ has been used in their compilation. For the spectral approach the height H corresponding to λ would be given by an empirical spectrum; in this case (4.8) should be used.

Example 4.1

If we are considering the wave forces on the columns of a concrete gravity structure (top line of Figure 4.7) we are in the 10 metre region.

● For a wave of height 1 metre we are in the diffraction regime.
● For a wave of height 5 metres the inertia forces dominate and we may possibly use the Morison equation below, with just the inertia term.
● For a wave of height 10 metres the drag forces are important and we must use the full Morison equation. We see that we are also in the post-critical region, where vortex shedding may occur, so we may have to consider lift forces in this case.

Morison's equation for a stationary slender member

In 1950 J. R. Morison and his associates[2], as a result of their experiments, postulated an empirical formula for the forces on vertical cylinders (not necessarily circular) in the presence of surface waves. They postulated that the force had two components:

● *the drag force*, proportional to the square of the water-particle velocity, the constant of proportionality being a drag coefficient having substantially the same value as for steady flow;

● *an inertial or virtual-mass force*, proportional to the horizontal component of acceleration of the water particles.

To these two components we add a lift component, if the conditions demand it. All water velocities and accelerations are evaluated at the axis of the cylinder *as if the cylinder were absent* using linear or Airy wave theory.

Note. It has been shown that the assumption regarding the first component above may not be entirely correct. Laird[9] has shown that, when the natural vibration frequency of a member approaches the eddy-shedding frequency, both the lift and drag forces are up to 4.5 times the drag force on a fixed circular cylinder in steady flow.

Considering for the moment drag and inertia forces, we may write the force per unit length in the x direction at depth z as:

$$\overset{\text{Inertia} \quad \text{Drag}}{F(z, t) = C_I \dot{v}_x + C_D v_x |v_x|} = F_I + F_D \tag{4.9}$$

This force is in the direction of wave advance, and the water-particle velocities and accelerations v and $\dot{v}_x$ are evaluated at the cylinder axis. C_I is a constant due to inertia consisting of two terms, one due to the 'hydrodynamic' mass contribution and the other to the variation of the pressure gradient within the accelerating fluid (as explained in section 4.1).

$$C_I = C_M + C_A = c_m \frac{\rho \pi D^2}{4} + \rho A = c_i \frac{\rho \pi D^2}{4} \tag{4.10}$$

where c_i is the inertia coefficient for the section, c_m is the added mass coefficient, and A is the cross-sectional area. Hence the inertia force on the body per unit length can be written:

$$F_I(z, t) = C_I \dot{v}_x = C_M \dot{v}_x + C_A \dot{v}_x$$

$$= c_m \frac{\rho \pi D^2}{4} \dot{v}_x + \rho A \dot{v}_x \tag{4.11}$$

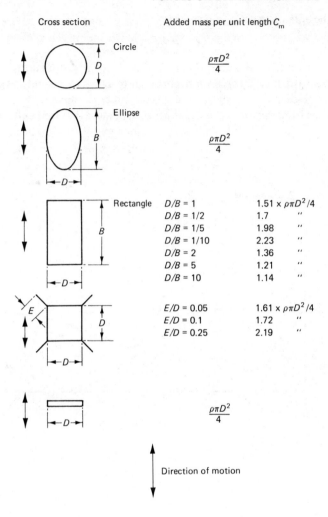

Circle

$$\frac{\rho\pi D^2}{4}$$

Ellipse

$$\frac{\rho\pi D^2}{4}$$

Rectangle

$D/B = 1$	$1.51 \times \rho\pi D^2/4$
$D/B = 1/2$	1.7 "
$D/B = 1/5$	1.98 "
$D/B = 1/10$	2.23 "
$D/B = 2$	1.36 "
$D/B = 5$	1.21 "
$D/B = 10$	1.14 "

$E/D = 0.05$	$1.61 \times \rho\pi D^2/4$
$E/D = 0.1$	1.72 "
$E/D = 0.25$	2.19 "

$$\frac{\rho\pi D^2}{4}$$

Direction of motion

Figure 4.8   Inertia coefficients

In Figure 4.8 inertia coefficients for some sections are presented in relation to the coefficient for an infinite circular cylinder in an infinite medium, $c_m = 1$. Drag coefficients are more difficult to find in the references, but those for a circular cylinder are shown in Figure 4.9 as a function of Reynolds number. The relationship between the coefficients c_d and C_D is given by:

$$C_D = \tfrac{1}{2} c_d \rho D \qquad (4.12)$$

We can see from the above that the drag force is proportional to D whereas the inertia force is proportional to D^2. As we would expect, for large-diameter members the inertia forces dominate.

It has been shown by Keulegan and Carpenter[5] that the coefficients

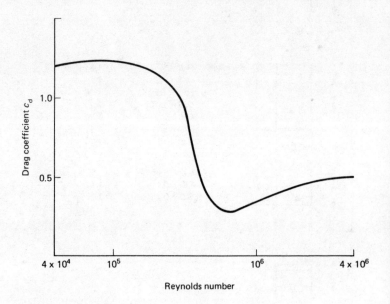

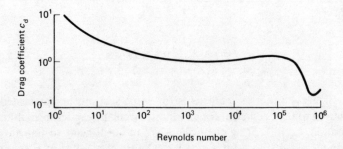

Figure 4.9   Drag coefficient as a function of Reynolds number, for a circular cylinder

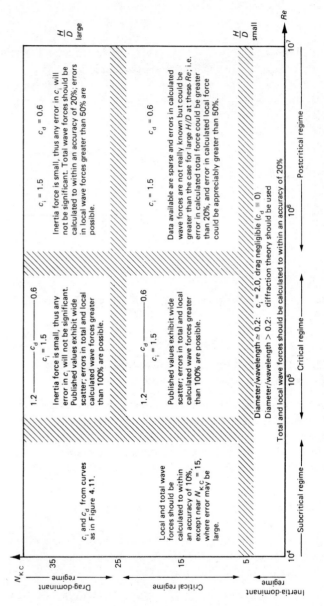

Figure 4.10 c_i *and* c_d *as functions of* Re *and* N_{KC}, *appraised from the literature for smooth, vertical, surface-piercing cylinders (from BSRA Report[17])*

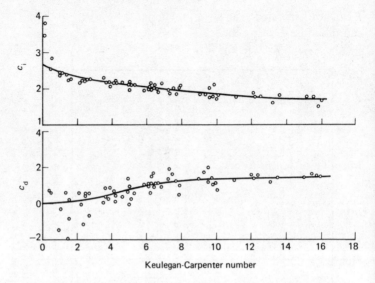

Figure 4.11 c_i and c_d as functions of N_{KC} only (from Chakrabarti et al.[10])

c_i, c_m and c_d depend also on the Keulegan–Carpenter number N_{KC} defined by:

$$N_{KC} = V_m T/D \qquad (4.13)$$

where V_m is the maximum horizontal water-particle velocity
T is the time period of the wave

This number N_{KC} is important when we come to consider lift forces at the end of this section. Figure 4.10, from reference 17, shows a summary of the expected values of c_i and c_d as functions of Re and N_{KC}. Figure 4.11, from Chakrabarti[10], shows the variation of c_i and c_d with Keulegan–Carpenter number only, for a circular cylinder.

Linearisation of the Morison equation

The drag term in equation (4.10) is non-linear and a linearisation is now necessary to solve the problem. If we assume that a linearised drag coefficient $\overline{C}_D$ exists[17], we can write:

$$F = C_M \dot{v}_x + \overline{C}_D v_x + C_A \dot{v}_x + C_D v_x |v_x| - \overline{C}_D v_x$$
$$= C_M \dot{v}_x + \overline{C}_D v_x + C_A \dot{v}_x + E \qquad (4.14)$$

E can be interpreted as an error function that we try to make zero. We can minimise E using the least-square technique.

$$\left\langle \frac{\partial E^2}{\partial \overline{C}_D} \right\rangle = -2\langle (C_D v_x^2 |v_x| - \overline{C}_D v_x^2 \rangle = 0 \tag{4.15}$$

$\langle \ \rangle$ denotes the time average. From (4.15) we obtain:

$$\overline{C}_D = C_D \frac{\langle v_x^2 |v_x| \rangle}{\langle v_x^2 \rangle} \tag{4.16}$$

For a Gaussian process with a zero mean, one has[12]:

$$\langle v_x^2 \rangle = \sigma_{v_x}^2$$
$$\langle |v_x| \rangle = \sqrt{(8/\pi)}\sigma_{v_x} \tag{4.17}$$
$$\langle v_x^2 |v_x| \rangle = \sqrt{(8/\pi)}\sigma_{v_x}^3$$

Hence we obtain:

$$\overline{C}_D = C_D \sqrt{(8/\pi)}\sigma_{v_x} \tag{4.18}$$

Note that to obtain equation (4.18) we need to know the distribution of v, the wave-particle velocities. Here we have assumed that we have a Gaussian distribution for the surface elevation η and hence a Gaussian distribution for v. These assumptions are discussed fully in Chapter 3. The linearised Morison equation can now be written:

$$F(z, t) = C_I \dot{v}_x + C_D \sqrt{(8/\pi)}\sigma_{v_x} v_x \tag{4.19}$$

Inclined members

So far we have only considered vertical cylinders in our calculations. Many structures, particularly of the steel lattice type, will contain a number of bracing members that are at an angle to the vertical and sufficiently near to the surface to be affected by wave motion.

In the conventional Morison equation we see that the (horizontal) force involves only the horizontal components of the fluid velocity and acceleration, i.e. the component of motion perpendicular to the axis of the member. In fact the fluid velocity and acceleration vectors will be at an angle to the horizontal depending on the phase of that part of the wave we think of as coinciding with the axis. The tangential components of these vectors are neglected in the calculation; although tangential components have a small effect on the member, owing to skin friction, typically the coefficients that would represent this effect would be 30 to 120 times smaller than the conventional drag and inertia coefficients.

There is no general agreement on how the Morison equation should be extended to deal with inclined members. Four different formulations are discussed by Wade and Dwyer[18]. The most conservative method involves using the moduli of the velocity and acceleration vectors in the Morison equation (in place of the horizontal components usually used) for situations in which the angle between the member's axis and these vectors is less than 60°. This method is recommended by Ippen[19]. This method is, however, not consistent with the original Morison equation for a vertical member. Other methods involve coefficients for elliptical cross-sections presented to the flow and various assumptions about pressures normal to the member.

Wade and Dwyer[18] demonstrated that the different approaches resulted in a 22 per cent variation of maximum base shear for a four-pile structure examined, and a 17 per cent variation for an eight-pile structure. The difference between the methods becomes more important as drag predominates.

Here we shall follow the method proposed by Borgman[20] and improved by Chakrabarti[21]. This method is one of the least conservative but is conceptually consistent with the classical Morison equation for a vertical member, which can be seen to be a special case. If we write the Morison equation (4.9) in vectorial form, we obtain for the force per unit length at depth z the general expression:

$$\vec{F}(z, t) = C_I \dot{\vec{v}}_n + C_D \vec{v}_n |\vec{v}_n| \tag{4.20}$$

where $\vec{v}_n$ and $\dot{\vec{v}}_n$ are interpreted as the normal water-particle velocity and acceleration vectors, and $| \, |$ represents the vector modulus. For the case of a vertical cylinder, $\vec{v}_n$ would be replaced simply by v_x for a wave travelling in the positive x direction, given equation (4.9).

If, following Chakrabarti, we take $\vec{c}$ to be a unit vector along the cylinder axis (see Figure 4.12) and resolve the water-particle velocities in two perpendicular directions $\vec{v}_t$ and $\vec{v}_n$ (where $\vec{v}_t$ is the component parallel to $\vec{c}$, and $\vec{v}_n$ is the component perpendicular to $\vec{c}$ and in the plane of $\vec{v}_t$ and $\vec{c}$), we then have:

$$\vec{v}_t = (\vec{v} . \vec{c})\vec{c} \tag{4.21}$$

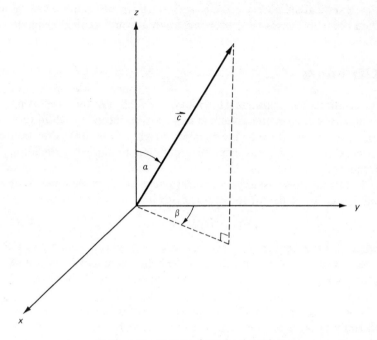

Figure 4.12 Notation for inclined member

but:
$$\vec{v} = \vec{v}_n + \vec{v}_t \qquad (4.22)$$

$\therefore$
$$\vec{v}_n = \vec{v} - (\vec{v}.\vec{c})\vec{c} \qquad (4.23)$$

As the wave is taken to be travelling along the positive x axis, $\vec{v}$ has no y component. We may therefore write equation (4.23) in component form as:

$$v_{nx} = v_x - (v_x c_x + v_z c_z)c_x$$
$$v_{ny} = -(v_x c_x + v_z c_z)c_y \qquad (4.24)$$
$$v_{nz} = v_z - (v_x c_x + v_z c_z)c_z$$

In terms of the angles defined in Figure 4.12, the components of $\vec{c}$ may be written:

$$c_x = \sin\alpha\cos\beta$$
$$c_y = \sin\alpha\sin\beta \qquad (4.25)$$
$$c_z = \cos\alpha$$

For a vertical cylinder $\alpha = 0$ and $v_n = (v_x, 0, 0)$, and equation (4.20) then becomes the classical Morison equation for a vertical cylinder.

Lift forces

As mentioned in section 4.1, the motion of fluid past the supporting members of a structure gives rise to vortex shedding, which in turn gives rise to a lift force on the member at right angles to the direction of fluid flow at a frequency corresponding to the eddy-shedding frequency.

For steady flow past a cylinder due to a current the main measure of the frequency of the lift forces is the Strouhal number S, defined by:

$$S = Df/V \tag{4.26}$$

where f is the frequency of vortex shedding. Figure 4.13 is a plot of S against Re for a circular cylinder. The higher values of Re of course correspond to faster flow. More will be said about this topic in Chapter 6.

For slender members we may expect vortex shedding to occur during the wave motion. The situation, however, is more complex than the one considered above for currents, because the flow is

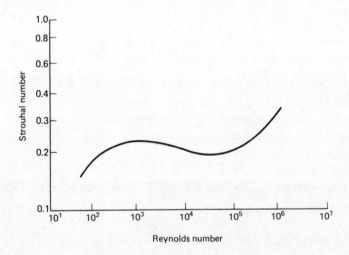

Figure 4.13   Strouhal number for vortex shedding behind circular cylinder

accelerating and reverses direction regularly. In many cases the flow does not persist long enough in one direction for a 'vortex street' or wake to develop before flow reversal. The disturbance of the fluid developed in one half-cycle of the wave will be swept back on to the member in the following half-cycle, becoming superimposed on the approaching flow. For vortex shedding to occur we would expect (considering the acceleration from rest of a circular cylinder) that the orbital velocity of the water particles should stay sensibly constant for three to four times the vortex-shedding frequency. So dynamic lift forces are only likely to occur for long waves and small cylinders. We should also bear in mind that the water-particle velocity decreases with depth, so that flow separation and hence vortex shedding will only occur at the surface, if at all. Hence the lift forces in waves are confined to the surface, but with currents they extend often to the sea bed.

In the present context we are considering an accelerating flow, and a more appropriate measure of the flow regime in which lift forces will occur is the Keulegan–Carpenter number N_{KC}, already defined in equation (4.13). Keulegan and Carpenter[5] proposed that when $N_{KC} \geqslant 15$ lift forces will occur. Figure 4.14 (from Bidde[6]) shows the development of vortex wakes in oscillating flow and gives the corresponding Keulegan–Carpenter numbers. We see from the diagrams that wakes are apparently being formed for Keulegan–Carpenter numbers less than 15. This conclusion must be suspect, however, because the observations were purely visual, and arguments in Shaw[7] considering the orbit widths of certain waves tend to throw more doubt on Bidde's conclusions.

Bidde also tried to take into account the effect that the wave height has on the relative importance of lift and longitudinal forces (drag and inertia). Figure 4.15 gives some idea of this effect for a pile of diameter 0.5 ft (150 mm) for waves of differing time period.

As a rough guide to the above we may use the following criteria, first proposed by Hogben[8]:

$D/\lambda > 1$ Conditions approximate to pure reflection
$D/\lambda > 0.2$ Diffraction increasingly important
$D/w_o > 0.2$ Inertia increasingly predominant
$D/w_o < 0.6$ Incipience of lift and drag
$D/w_o < 0.2$ Drag increasingly predominant

where w_o is the orbit width parameter of the water-particle motion. In

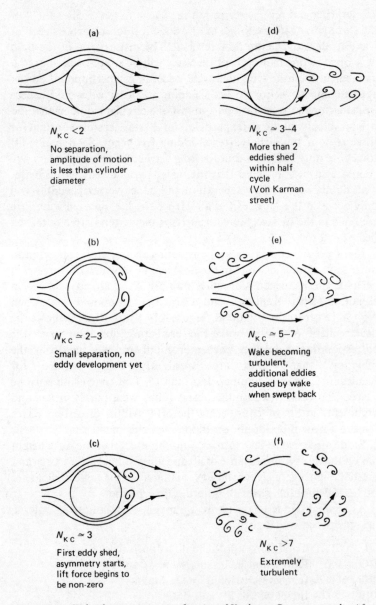

(a)

$N_{KC} < 2$

No separation,
amplitude of motion
is less than cylinder
diameter

(d)

$N_{KC} \simeq 3-4$

More than 2
eddies shed
within half
cycle
(Von Karman
street)

(b)

$N_{KC} \simeq 2-3$

Small separation, no
eddy development yet

(e)

$N_{KC} \simeq 5-7$

Wake becoming
turbulent,
additional eddies
caused by wake
when swept back

(c)

$N_{KC} \simeq 3$

First eddy shed,
asymmetry starts,
lift force begins to
be non-zero

(f)

$N_{KC} > 7$

Extremely
turbulent

Figure 4.14 *Wake characteristics as a function of Keulegan–Carpenter number (from Bidde[6])*

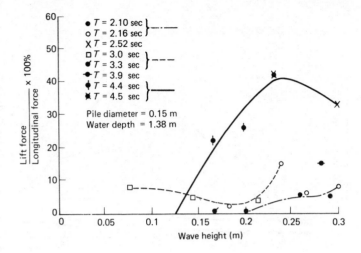

Figure 4.15   Wave height versus ratio of lift to longitudinal force (from Bidde[6])

the case of waves in deep water w_0 is equal to the wave height H at the surface.

If we suspect that vortex shedding may occur in the domain of our problem, then as a first approximation we may postulate that the force component due to lift F_L would be of the form:

$$F_L(z, t) = C_L v_x |v_x| \qquad (4.27)$$

where $C_L = \frac{1}{2}\rho D c_l$ (since the force has the nature of a drag force)

F_L is now perpendicular to the direction of wave advance (the x direction)

c_l is a lift coefficient

However, it is observed (by Chakrabarti[11] in particular) that the frequency of the lift force is always some multiple of the wave frequency, so equation (4.27) seems inadequate to describe this behaviour.

A water-particle velocity v with harmonic time dependence $\exp(i\omega t)$ would give a lift force with time dependence $\exp(2i\omega t)$, i.e. with twice the wave frequency. As a result of this, Chakrabarti proposed the following form for the lift force:

$$F_L(z, t) = \frac{1}{2}\rho D V_m^2 \sum_{n=1}^{N} c_l^n \cos(n\omega_0 t + \psi_n) \qquad (4.28)$$

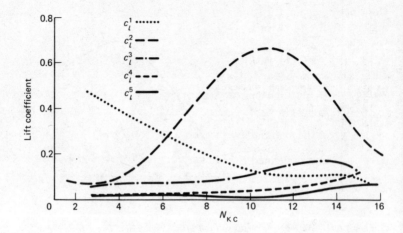

Figure 4.16   First five coefficients of lift force, obtained by harmonic analysis (from Chakrabarti et al.[10])

where V_m and D are as before

ω_o is the angular frequency of the incident wave

ψ_n is the phase angle of the nth harmonic force (corresponding to a force with frequency n times that of the incident wave)

c_{ℓ}^n is the lift coefficient for the nth harmonic and is a function of N_{KC}, the Keulegan–Carpenter number

Figure 4.16 shows how the first five lift coefficients $c_{\ell}^1, c_{\ell}^2, c_{\ell}^3, c_{\ell}^4$ and c_{ℓ}^5 vary with N_{KC} obtained from the test results on a 3 inch (76 mm) diameter circular cylinder. We can see that c_{ℓ}^2 (the coefficient corresponding to the component with frequency equal to twice the wave frequency) dominates the others for most of the range of N_{KC} considered. For $N_{KC} \simeq 15$, the total horizontal force magnitude given by

$$F_R = \sqrt{\{[C_I \dot{v}_x + C_D \sqrt{(8/\pi)} \sigma_{v_x} v_x]^2 + F_L^2\}} \qquad (4.29)$$

is 60 per cent higher than the longitudinal in-line force

$$F = C_I \dot{v}_x + C_D \sqrt{(8/\pi)} \sigma_{v_x} v_x \qquad (4.30)$$

given by the normal form of the Morison equation.

4.3 WAVE FORCES ON MOVING MEMBERS

Considering again, for the moment, only drag and inertia forces, we may generalise the Morison equation (4.9) for a moving cylinder as:

$$\vec{F}(z, t) = C_M \frac{D\vec{r}}{Dt} + C_A \frac{D\vec{v}}{Dt} + C_D \vec{r}|\vec{r}| \tag{4.31}$$

where $\vec{r}$ is the relative velocity of the fluid $\vec{v}$ and the structure $\vec{u}$, i.e. $\vec{r} = \vec{v} - \vec{u}$

$\dfrac{D}{Dt}$ is the material derivative following the motion of the fluid,

and is defined by:

$$\frac{D}{Dt} = \frac{\partial}{\partial t} + \vec{v}.\nabla \tag{4.32}$$

We have written the material derivative here because the variables are evaluated at the moving axis of the cylinder. As we are considering only a linear wave theory, however, we may immediately neglect the convective acceleration and approximate the material derivative by the local one, $\partial/\partial t$.

For simplicity, consider motion of the cylinder in the direction of the wave motion; the members would be expected to move in this way in any case, in the absence of vortex shedding. We then have only the scalars $r = v_x - \dot{u}$ to consider, so we write (4.31) as:

$$F(z, t) = C_M \dot{r} + C_A \dot{v}_x + C_D r|r| \tag{4.33}$$

The first term is the added or hydrodynamic mass term, as before, and we see that it depends on the motion of the member under consideration. The second term is the inertial term representing the distortion of the streamlines in the fluid, and is independent of the acceleration of the structure in the linear approximation we are using here. The third term is the familiar drag term, which of course is proportional to the square of the relative velocity of the structure and fluid, since it is a viscous effect.

Equation (4.33) may be linearised in the same way as equation (4.14) to give:

$$F(z, t) = C_M \dot{r} + C_A \dot{v}_x + C_D \sqrt{(8/\pi)} \sigma_r r \tag{4.34}$$

This time, however, we are not justified in assuming that the distribution of r is Gaussian, as is necessary for this linearisation technique. We can start the problem with the σ_{v_x} corresponding to the velocities for a rigid pile (a Gaussian distribution) and use a cyclic procedure to obtain σ_r. Using this procedure, convergence of the solution can be assured[12]. It is important to note that the use of the initial σ_{v_x} in the calculations, without any cyclic improvement to obtain σ_r, can give place to errors in analysis of offshore frame structures, for which the drag effects are not negligible when compared with the inertia effects.

The above analysis may be found in Brebbia *et al.*[13].

Lift forces for a moving cylinder

If we try to extend the expression for the lift force (4.28) to the case of a moving cylinder, we may no longer assume that the motion will be in the direction as the wave field; hence we must use the full vector form for the relative velocity $\vec{r} = \vec{v} - \dot{\vec{u}}$. Then we may define r_m, the magnitude of the relative velocity, by:

$$r_m = \max\left(|\vec{v} - \dot{\vec{u}}|\right) \tag{4.35}$$

where $|\ |$ here means the vector modulus.

As an analogy with the extension of the calculation of the drag forces to a moving member, we may extend the formulation for the lift forces in the same way to:

$$F_L(z, t) = \frac{1}{2}\rho D r_m^2 \sum_{n=1}^{N} c_r^n \cos\left(n\omega_0 t + \psi_n\right) \tag{4.36}$$

If we assume that r has a Gaussian probability distribution then r_m will have a Rayleigh distribution, so we must be careful if we are to attempt to linearise the above expression in an analogous way to the drag linearisation.

Example 4.2

We now apply the foregoing analysis to a vibrating slender column of circular cross-section in the presence of a two-dimensional Airy wave of frequency ω_0. We neglect the lift forces here to simplify the analysis so that we may use scalar variables for the relative velocities and

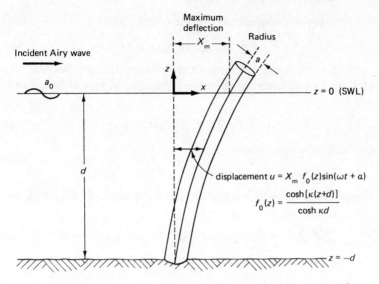

Figure 4.17   Flexible column vibrating in presence of linear wave

accelerations, because the column will vibrate in the same direction as the direction of wave propagation. We also assume that the column has a mode shape corresponding to the particle motions in an Airy wave, but with a much smaller amplitude (see Figure 4.17), and that it moves in a harmonic way with the frequency of the incident wave. This last is not an unreasonable assumption.

For an Airy wave of amplitude a_o we have (from Chapter 3):

$$\Phi = \frac{ga_o}{\omega} f_o(z) \sin(\omega t - \kappa x) \tag{a}$$

where $\omega^2 = g\kappa \tanh \kappa d$, and we have written $f_o(z)$ for the usual

$$\frac{\cosh\left[\kappa(z+d)\right]}{\cosh \kappa d}$$

factor. Hence:

$$v_x = \frac{\partial \Phi}{\partial x} = -\frac{\kappa g a_o}{\omega} f_o(z) \cos(\omega t - \kappa x) \tag{b}$$

and:

$$\dot{v}_x = \kappa g a_o f_o(z) \sin(\omega t - \kappa x) \tag{c}$$

We have for the displacement of the column axis:

$$u = X_m f_o(z) \sin(\omega t + \alpha) \tag{d}$$

where α is the angle representing the phase difference between the motion of the column and the wave. Hence the velocity of the column axis is:

$$\dot{u} = + \omega X_m f_o(z) \cos(\omega t + \alpha) \tag{e}$$

and its acceleration is given by:

$$\ddot{u} = - \omega^2 X_m f_o(z) \sin(\omega t + \alpha) \tag{f}$$

Hence, from Morison's equation, the force per unit length in the x direction at depth z is given by:

$$F(z, t) = C_M(\dot{v}_x - \ddot{u}) + C_A \dot{v}_x + C_D(v_x - \dot{u})|v_x - \dot{u}| \tag{g}$$

From equation (4.32), for a circular cylinder radius a, taking:

$$c_m = c_a = c_d = 1$$
$$C_M = \pi a^2 \rho$$
$$C_A = \pi a^2 \rho$$
$$C_D = \rho a$$

we obtain:

$$F(z, t) = \pi a^2 \rho f_o(z) \{2g\kappa a_o \sin(\omega t) + \omega^2 X_m \sin(\omega t + \alpha)\} \times$$

$$\times \rho a \left[f_o(z) \right]^2 \left[\frac{g\kappa a_o}{\omega} \cos(\omega t) + \omega X_m \cos(\omega t + \alpha) \right] \times$$

$$\times \left| \left\{ \frac{g\kappa a_o}{\omega} \cos(\omega t) + \omega X_m \cos(\omega t + \alpha) \right\} \right| \tag{h}$$

For deep water, $f_o(z) \simeq \exp(\kappa z)$ and $\omega^2 \simeq g\kappa$. Equation (h) becomes:

$$F(z, t) = \exp(\kappa z)\pi a^2 \rho\omega[2a_o \sin(\omega t) + X_m \sin(\omega t + \alpha)] -$$

$$- \exp(2\kappa z)a\rho\omega^2[a_o \cos(\omega t) + X_m \cos(\omega t + \alpha)] \times$$

$$\times |a_o \cos(\omega t) + X_m \cos(\omega t + \alpha)| \tag{i}$$

The total force on the cylinder $F_t(t)$ is obtained from F by integrating $F(z, t)$ from $z = -d$ to $z = 0$ in equation (e), i.e.

$$F_t(t) = \int_{-d}^{0} F(z, t)\,dz \qquad \text{(j)}$$

The moment M about the base of the column is obtained from:

$$M(t) = \int_{-d}^{0} (z + d)\,F(z, t)\,dz \qquad \text{(k)}$$

Alternatively the upper limit may be taken as $\eta(t)$, the surface elevation at time t at $x = 0$.

4.4 COMPUTATION OF INERTIA COEFFICIENTS

If in the above example we consider the case of a stationary cylinder, $X_m = 0$. Then F becomes:

$$F(z, t) = 2\pi a^2 \rho g \kappa a_o f_o(z) \sin \omega t + \rho a [f_o(z)]^2 \times$$
$$\times \frac{g^2 \kappa^2 a_o^2}{\omega^2} \cos \omega t |\cos \omega t| \qquad (4.37)$$

We see that the two parts of the force corresponding to drag and inertia are now out of phase by $90°$. Hence if we calculate the inertia force at, say, $t = \pi/2\omega$ the drag component will be zero, as $\cos \pi$ is zero, and the force will be completely due to inertia. If we calculate the force on an obstacle by another method neglecting viscosity, we obtain the inertia component and hence the inertia coefficient. This may now be used in the full Morison equation, and we may hence include drag effects in the analysis, assuming we have a drag coefficient.

In Chapter 5 we describe the diffraction theory method of calculation of wave forces on bodies in an inviscid fluid. This is one method of calculating the inertia coefficients, although its complexity means that the method is restricted to simple geometries. The other methods involve the numerical solution of the corresponding inviscid (potential) problems.

Let us now deduce the mass coefficients for the inertia forces acting on an arbitrary structure. These coefficients are then arranged in a matrix, which is sometimes called the hydrodynamic mass matrix.

The first study of this topic is due to Westergaard[14], who determined the hydrodynamic forces on dams during earthquakes. He made two main hypotheses: (a) incompressibility of the water and (b) absence of surface waves. The reservoir was considered to be infinite. Westergaard solved the problem by using an analytical solution for the pressure distribution over a rigid wall subjected to harmonic ground motion. The hydrodynamic forces produced are opposite in direction to the ground motion and are proportional to the acceleration. Hence they can be represented as an equivalent mass added to the mass of the dam. More recently there have been a series of experimental and numerical studies to determine equivalent masses for dams.

To obtain the governing equations for this problem for a two-dimensional case such as the one in Figure 4.18, we use the Navier–Stokes equations—see (3.1) and (3.8)—restated here for reference:

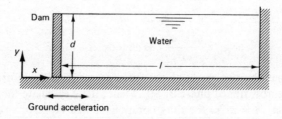

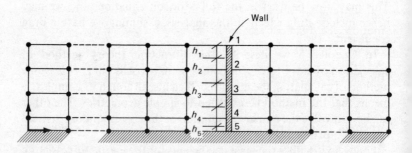

Figure 4.18   Fluid–structure systems: (a) dam moving horizontally in reservoir; (b) wall moving inside the water

$$\rho \, \frac{D\vec{v}}{Dt} = \rho \vec{F} - \vec{\nabla}p - \mu \vec{\nabla} \times \vec{\omega} \tag{4.38}$$

$$\vec{\nabla}.\vec{v} = 0 \tag{4.39}$$

We now make the following assumptions:
- The convective terms can be neglected for this type of motion.
- The fluid is inviscid, as well as incompressible.
- The only body force acting on the system is gravity.

Hence we have the two momentum equations:

$$\rho \, \frac{\partial v_x}{\partial t} = -\frac{\partial p}{\partial x}$$

$$\rho \, \frac{\partial v_y}{\partial t} = -\frac{\partial p}{\partial y} - g\rho \tag{4.40}$$

The continuity equation (4.39) can be differentiated with respect to time, which gives:

$$\frac{\partial^2 v_x}{\partial t \, \partial x} + \frac{\partial^2 v_y}{\partial t \, \partial y} = 0 \tag{4.41}$$

Substituting equations (4.40) into (4.41), we obtain the following Laplacian:

$$\nabla^2 P = \frac{\partial^2 P}{\partial x^2} + \frac{\partial^2 P}{\partial y^2} = 0 \tag{4.42}$$

where P is the total pressure and is equal to:

$$P = p + \rho g(d - y) \tag{4.43}$$

$\rho g y$ represents the hydrostatic pressure and p the pressure due to the motion, the dynamic pressure.

The boundary conditions corresponding to equation (4.42) are:
- There is no separation between the fluid and the structure. Hence on the face of the structure in contact with the fluid we have:

$$\partial P/\partial n = -\rho \ddot{u}_n \tag{4.44}$$

where $\ddot{u}_n$ is the normal component of acceleration at the interface.
- The surface waves are assumed to be of small amplitude and such that their effect on pressure can be neglected. Thus:

$$P = 0 \quad \text{on the free surface} \tag{4.45}$$

● At the bottom of the reservoir and sides, other than the structure face, the boundary condition is:

$$\partial P/\partial n = 0 \qquad (4.46)$$

Numerical Solution. Consider now the numerical solution of the problem shown in Figure 4.18. We can study the motion of each of the points 1 to 5 on the solid by assuming a unit acceleration for each of the h_i segments $(i = 1, \ldots 5)$. Then we determine the pressures at any of the points of the grid from the solution of Laplace's equation with boundary conditions (4.44) to (4.46). This can be done by using a numerical method (such as finite differences, finite elements or boundary integral equations). The forces at nodes 1 to 5 due to the pressures obtained after applying the unit accelerations can be interpreted as the coefficients of a square matrix $\mathbf{M_H}$, such that:

$$\mathbf{F_H} = \mathbf{M_H}\ddot{\mathbf{U}} \qquad (4.47)$$

where $\ddot{\mathbf{U}}$ is a vector of accelerations and $\mathbf{F_H}$ is a vector of nodal forces, for nodes 1 to 5.

In principle the hydrodynamic mass matrix can be obtained for any type of structure. In practice the limitations are the computer time and storage necessary for the solution of complex structures. When using a numerical technique we generally approximate the infinite domain by a finite one. This requires accepting that at a certain distance from the body we can assume that a solid boundary with boundary conditions of the type $\partial\phi/\partial n = 0$ exists. As an indication of the necessary length for this finite domain l, results for the problem shown in Figure 4.18(a) for a ratio $l/d = 4$ are in fair agreement with those obtained for $l/d = \infty$. Hence a ratio $l/d = 4$ in the numerical solution may be taken to represent $l/d \rightarrow \infty$.

A note of caution should be given about the theoretical inertia coefficients deduced by using potential theory. These coefficients are not strictly valid, since the viscous effects are not taken into account. Specifically, the wake behind the obstruction may significantly alter the results. The only practical solution in many cases is to use experimental coefficients when available.

4.5 EARTHQUAKE EFFECTS

As mentioned in section 4.1, many offshore structures are now being sited in locations that are seismically active, so the designer may have to take into account the effects of earthquakes on his structure.

The ground acceleration $\ddot{u}_g$ during the earthquake is usually assumed to be a zero mean ergodic Gaussian process. In this case the best method of analysis is to use a step-by-step analysis in the time domain using an actual record of an earthquake.

If the autocorrelation function of the acceleration record is reasonably small for time separations of 20–50 seconds, we may assume the process is stationary and use a spectral approach. Following most researchers in the subject (Kanai[15] and Tajimi[16]), we assume that the power spectral density of $\ddot{u}_g$ is given by:

$$S_{\ddot{u}_g \ddot{u}_g}(\omega) = \frac{|1 + 4\gamma_g^2(\omega/\omega_g)^2|S_o}{|1 - (\omega/\omega_g)^2|^2 + 4\gamma_g^2(\omega/\omega_g)^2} \qquad (4.48)$$

where S_o is a constant representing the strength of the disturbance
  ω_g is the characteristic ground frequency, 5–6π rad/s
  γ_g is the characteristic ground damping ratio for the location considered (typically 0.6–0.7 for a hard layer)

Once this spectrum is known, the drag, wave radiation damping and inertia effects of the water surrounding the structure must be taken into account.

Here we assume that the movement of the sea bed is horizontal and does not move the ocean appreciably. We also assume that we are dealing with slender members, which cannot produce waves of appreciable amplitude by their motion, so we may here neglect the effect of radiation damping. In Chapter 5 the topic of radiation damping is dealt with for large-diameter members using diffraction theory.

Consider now the motion of the supporting column of an offshore structure in isolation (Figure 4.19). The diagram is drawn from the point of view of a fixed observer; the sea bed has been displaced a distance u_g from its mean position, and the column has flexed so that its actual displacement from where the axis of a perfectly rigid column would be is a distance u at depth z. The total displacement of the part of the column considered is then given by:

$$u_t = u_g + u \qquad (4.49)$$

Now from the point of view of the stiffness of the column the displacement of the column is u, hence we write Ku for this term.

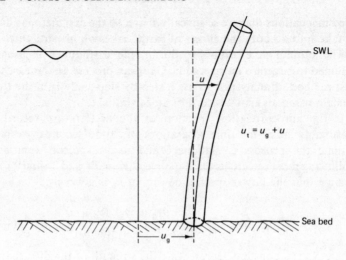

Figure 4.19   Deflection of single column during earthquake

Similarly the structural damping term is given by $C\dot{u}$ (since velocities are relative). The inertia term for the structure must, however, take into account the full motion of the column, since accelerations are not relative, and the column will feel a body force because of the ground acceleration. We therefore write $M\ddot{u}_t$ for this term, where M is the mass of this section of the column. Hence the equation of motion of this part of the column is:

$$M\ddot{u}_t + C\dot{u} + Ku = F(z, t) \qquad (4.50)$$

where $F(z, t)$ is the usual Morison force for a moving structural member. The concepts of stiffness and structural damping will be clarified in Chapter 7, where the response of a one-degree-of-freedom system is discussed in full.

$F(z, t)$ is given by the linearised form of the Morison equation:

$$F(z, t) = C_M(\dot{v} - \ddot{u}_t) + C_A\dot{v} + C_D\sqrt{(8/\pi)}\,\sigma_r(v - \dot{u}_t) \qquad (4.51)$$

where r is the relative velocity of the structure and fluid, i.e.

$$r = v - \dot{u}_t \qquad (4.52)$$

and σ_r is taken to be equal to σ_v as a first approximation. We see that the hydrodynamic mass term $C_M(\dot{v} - \ddot{u}_t)$ has to take into account the total motion, but the inertia term (which is independent of the motion of the structure) contains only the fluid acceleration $\dot{v}$ as we now expect. The (linearised) drag term $C_D\sqrt{(8/\pi)}\,\sigma_r(v - \dot{u}_t)$ must also take into account the total relative motion, as before. Rewriting equation (4.51) in terms of the variables u, u_g and v, we obtain, making our approximation for σ_r, the equation:

$$M(\ddot{u} + \ddot{u}_g) + C\dot{u} + Ku = C_M[\dot{v} - (\ddot{u} + \ddot{u}_g)] + C_A\dot{v} +$$
$$+ C_D\sqrt{(8/\pi)}\sigma_v[v - (\dot{u} + \dot{u}_g)] \quad (4.53)$$

Collecting together damping and inertia terms, we obtain:

$$(M + C_M)\ddot{u} + (C + C_D\sqrt{(8/\pi)}\sigma_v)\dot{u} + \kappa u$$
$$= -(M + C_M)\ddot{u}_g - C_D\sqrt{(8/\pi)}\,\sigma_v\dot{u}_g + (C_M + C_A)\dot{v} + C_D\sqrt{(8/\pi)}\sigma_v v \quad (4.54)$$

Treating the variables $\ddot{u}_g$ and v as known functions, given the coefficients, we may then solve this problem for the displacement u. In fact we use a statistical approach, as only the statistical nature of $\ddot{u}_g$ is known.

References

1. Penzien, J., Maharaj, K. K., and Berge, B., 'Stochastic response of offshore towers to random sea waves and strong motion earthquakes', in *Computers and structures*, Vol. 2, 733–756, Pergamon Press (1972)
2. Morison, J. R., O'Brien, M. P., Johnson, J. W., and Schaaf, S. A., 'The force exerted by surface waves on piles', *Petrol. Trans.*, *AIME*, **189** (1950)
3. Muir Wood, A. M., *Coastal hydraulics*, 136, Macmillan (1969)
4. Hogben, N., 'Wave loads on structures', Proc. Behaviour of Offshore Structures Conf., Trondheim (1976)
5. Keulegan, G. H., and Carpenter, L. H., 'Forces on cylinders and plates in an oscillating fluid', *J. Res. Natn. Bur. Stand.*, **60**, No. 5, 423–440 (1958)
6. Bidde, D. D., Tech. Report HEL 9–16, Hydraulic Engineering Laboratory, Univ. California, Berkeley (1970)
7. Shaw, T. L., 'Wave induced loading on dynamic structures', Conf. on Offshore Structures, Instn Civil Engrs, London (1974)
8. Hogben, N., *Fluid loading of offshore structures, a state of art appraisal: Wave loads*, R. Inst. Naval Arch. (1974)
9. Laird, A. D. K., 'Water forces on flexible oscillating cylinders', *Proc. ASCE (Waterways and Harbors Div.)*, **88**, WW3 (Aug. 1962)
10. Chakrabarti, S. K., Wolbert, A. L., and Tam, W. A., 'Wave forces on vertical circular cylinder', *Proc. ASCE (Waterways, Harbors and Coastal Engng Div.)*, **102** (May 1976)
11. Malhotra, A. K., and Penzien, J., 'Non-deterministic analysis of offshore structures', *Proc. ASCE (Engng Mech. Div.)*, **96** (Dec. 1970)
12. Malhotra, A. K., and Penzien, J., 'Response of offshore structures to random wave forces', *Proc. ASCE (Struct. Div.)*, **96**, ST10 (Oct. 1970)

13. Brebbia C. A., *et al., Vibrations of engineering structures*, Southampton University Press (1974)
14. Westergaard, H. M., 'Water pressure on dams during earthquakes', *Trans. ASCE*, **98**, 418–433 (1933)
15. Kanai, K., 'Semi-empirical formula for the seismic characteristics of the ground', *Bull. Earthquake Res. Inst. Univ. Tokyo*, **35**, 308–325 (1957)
16. Tajimi, H., 'A structural method of determining the maximum response of a building structure during an earthquake', Proc. 2nd World Conf. Earthquake Engng, Tokyo and Kyoto, Vol. 11 (1960)
17. British Ship Research Association, *A critical evaluation of the data on wave force coefficients*, OSFLAG Project 10, Report No. W278, 2 Vols (1976)
18. Wade, B. G., and Dwyer, M., 'On the application of Morison's equation to fixed offshore platforms', OTC 2723, Proc. Offshore Technology Conf., Houston (1976)
19. Ippen, A. T., *Estuary and coastline hydrodynamics*, 341–375, McGraw-Hill (1966)
20. Borgman, L. E., 'Computation of the ocean-wave forces on inclined cylinders', *J. Geophys. Res., Trans. AGV*, **39**, No. 5 (Oct. 1958)
21. Chakrabarti, S. K., Tam, W. A., and Wolbert, A. L., 'Wave forces on a randomly oriented tube', OTC 2190, Proc. Offshore Technology Conf., Houston (1975)

5 Diffraction Problems

5.1 INTRODUCTION

In Chapter 4 we described a method of calculating the wave forces on the slender members of an offshore structure. In this context a slender member is taken to be one whose presence does not disturb the incident wave field appreciably. Such members are often used in the construction of steel structures.

There is, however, a new kind of structure that has proved more suitable in some situations: the concrete gravity structure, held in position against the environmental loads largely by its enormous weight. This solution to the problem of storage and exploitation of oil reserves has three main drawbacks:

● The volume of water displaced by the structure is extremely large, giving rise to large inertial forces.

● The structure presents a large profile to incident wave fields, necessitating a diffraction theory approach to the calculation of wave forces.

● The loads on the underlying sea bed are large.

Fortunately, from the point of view of calculation of the wave forces, the first of these problems puts us in the inertial regime, described by Hogben[1] as the regime in which

$$d/a_o > 0.4 \qquad (5.1)$$

where d is a typical dimension of the structure near the still water level and a_o is half the orbit width parameter for the water particles; in the case of a linear harmonic wave this can be taken as the wave

145

amplitude. In this chapter we therefore neglect drag forces due to the wave motion and consider only potential flow.

The geometry of a gravity structure consists typically of a large, approximately cylindrical, base surmounted by a number of cylindrical columns that intersect the water surface and support a platform on which the drilling equipment is placed. This simple geometry makes a diffraction theory analysis of the wave forces possible.

In this chapter a full stochastic, dynamic analysis of the wave forces is attempted. The method consists of first calculating the wave forces for an incident wave of fixed frequency and then using the 'direct transfer function method' to find the spectral density of the forces on the structure. The method is presented in full for the case of a single stationary vertical cylinder. These force spectra can then be used in the random vibration analysis of the response of offshore structures.

One difficulty with carrying out the full stochastic analysis is that the diffraction effects discussed are dependent on frequency; therefore no frequency-independent diffraction coefficients can be used in the analysis unless a cut-off frequency is chosen and the input spectra are considered to be zero beyond this frequency. Even for a single frequency a diffraction coefficient is difficult to find, since the *functional form* of the diffracted field is often completely different from the incident field, and so a simple ratio of variables derived from each field need not be independent of frequency or the dimensions of the obstacle. It is only in the regime where conditions approximate to pure reflection—described by Hogben[1] as the regime in which

$$d/\lambda > 1 \qquad\qquad (5.2)$$

(where λ is the wavelength of the incident wave)—that the functional form of the diffracted field is comparable with the incident one. In this case a reflection coefficient would be more appropriate.

Diffraction coefficients

Hogben and Standing[2] have, however, postulated some functional forms for diffraction coefficients as functions of the frequency ω or wavenumber κ. The coefficients c_h, c_v and c_m for circular cylinders of various aspect ratios, i.e. the height to diameter ratio $h/2a$, were defined as follows:

c_h = ratio of maximum total horizontal force to maximum Froude–Krylov force. The Froude–Krylov force would be the inertia force calculated from Morison's equation—see equation (4.9)—assuming the fluid motion to be undisturbed by the presence of the body.

c_v = ratio of maximum total to maximum Froude–Krylov vertical force.

c_m = the corresponding ratio for the overturning moments.

These coefficients were calculated for:

$$h/d < 0.6 \tag{5.3}$$

where d is the depth of water (the cylinders were standing on the sea bed) and:

$$0.3 < h/2a < 2.3 \quad \text{for } c_h \text{ and } c_v \tag{5.4}$$

$$0.6 < h/2a < 2.3 \quad \text{for } c_m \tag{5.5}$$

The approximate formulae for the coefficients under these circumstances are then:

$$c_h = 1 + 0.75(h/2a)^{1/3}\,(1 - 0.3\kappa^2 a^2) \tag{5.6}$$

$$c_v = 1 + 0.74\kappa^2 a^2 (h/2a) \quad \text{for } 1.48\kappa a(h/2a) < 1 \tag{5.7}$$

$$= 1 + 0.5\kappa a \quad \text{for } 1.48\kappa a(h/2a) > 1$$

$$c_m = 1.9 - 0.35\kappa a \tag{5.8}$$

The use of these coefficients entails only an error of up to 5 per cent as compared with the full diffraction theory.

For a stationary circular cylinder, therefore, the diffraction effects may be taken into account by writing the Morison equation (4.9) as:

$$F(z, t) = c_h [C_I \dot{v}_x] + C_D v_x |v_x| \tag{5.9}$$

This formulation does not, however, take into account interaction effects between the large columns of a concrete gravity structure, and it is inadequate to determine the forces on surface-piercing members. A full diffraction theory analysis will be used for these members in the diffraction regime, defined roughly by:

$$d/\lambda > 0.2 \tag{5.10}$$

Some idea of the effects to be taken into account in this chapter may

148

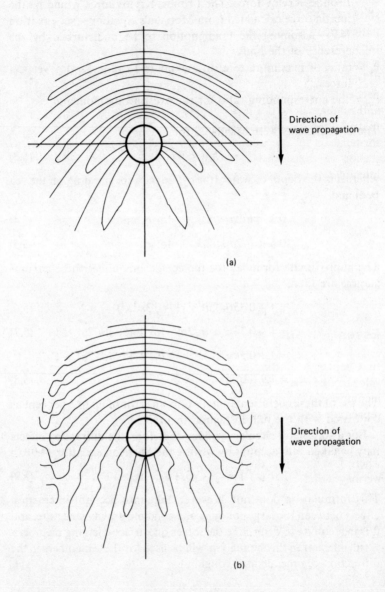

Direction of
wave propagation

(a)

Direction of
wave propagation

(b)

*Figure 5.1   Wave pattern around circular cylinder: (a) $\kappa a = 1.4$; (b) $\kappa a = 2\pi a/\lambda = 4$
(from Oortmerssen[3])*

be obtained by referring to Figure 5.1, taken from Oortmerssen[3]. In this figure the wave pattern around a vertical circular cylinder is shown for two values of κa. These may be interpreted as the wave patterns for two frequencies ω (or corresponding wavenumbers κ), or the wave patterns for two cylinders of differing radii in a wave field of a fixed frequency.

In the rest of this chapter we shall review the necessary wave theory needed for diffraction problems, examine the wave forces on columns and submerged vertical cylinders, and then try to examine the effect that column flexibility has on the wave forces. A finite element formulation for the solution of diffraction problems is presented in section 5.6, and other methods such as the Green's function and integral equation methods are discussed briefly.

5.2 WAVE THEORY FOR DIFFRACTION PROBLEMS

In this section we review some of the mathematics and fluid dynamics necessary for the solution of diffraction problems.

Essential fluid dynamics

In Chapter 3, starting from the Navier–Stokes equation, we developed the linear wave theory for the motion of an inviscid, incompressible fluid, of constant depth d, in irrotational motion. We reproduce the main results below for reference. We have:

$$\nabla^2 \Phi = 0 \qquad (5.11)$$

where $\Phi \equiv \Phi(x, y, z, t)$ is the velocity potential. For a fluid of depth d we may write:

$$\Phi(x, y, z, t) = \phi(x, y) f(z) \exp(+i\omega t) \qquad (5.12)$$

where $\phi(x, y)$ is the reduced velocity potential and

$$f(z) = \frac{\cosh[\kappa(z + d)]}{\cosh \kappa d} \qquad (5.13)$$

The reduced velocity potential now satisfies the Helmholtz equation in two dimensions:

$$\nabla^2 \phi + \kappa^2 \phi = 0 \tag{5.14}$$

where ∇^2 is now $(\partial^2/\partial x^2) + (\partial^2/\partial y^2)$, the two-dimensional Laplacian, and κ is given by the dispersion relation:

$$\omega^2 = g\kappa \tanh \kappa d \tag{5.15}$$

For a linear (two-dimensional) wave of frequency ω and amplitude a_0 we have:

$$\Phi(x, z, t) = \frac{ig}{\omega} a_0 \exp[-i(\omega t - \kappa x)] \frac{\cosh[\kappa(z+d)]}{\cosh \kappa d} \tag{5.16}$$

for a wave travelling in the positive x direction of wavelength

$$\lambda = 2\pi/\kappa \tag{5.17}$$

and celerity

$$c = \omega/\kappa \tag{5.18}$$

Diffraction theory method

In a diffraction theory analysis of the wave forces on a stationary body of arbitrary cross-section (considered for the moment to be vertical with symmetry in the z direction), we consider a linear plane wave of amplitude a_0 and frequency ω impinging on the body that penetrates the water surface and reaches to the sea bed, which is considered to be flat, impermeable and at a depth d. See Figure 5.2. In addition to the

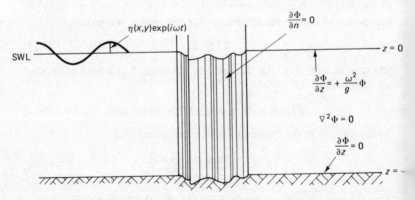

Figure 5.2 Boundary conditions and notation for diffraction problems

boundary conditions on the surface and sea bed we have to apply a boundary condition on the body surface. Since the body is impermeable there will be no normal flux of liquid through its surface and hence no normal velocity v_n. We write this:

$$v_n = \frac{\partial \Phi}{\partial n} = 0 \quad \text{on } S \text{ the surface of the body} \qquad (5.19)$$

where n is a local coordinate measured into the fluid. We may rewrite (5.19) as:

$$\vec{\nabla}\phi \cdot \vec{n} = \frac{\partial \phi}{\partial n} = 0 \quad \text{on } S \qquad (5.20)$$

The equation and boundary conditions are now all posed in terms of the reduced velocity potential $\phi(x, y)$, and we have a two-dimensional problem.

Now, as is general practice in diffraction theory, we split the reduced potential $\phi(x, y)$ into two parts: $\phi_i(x, y)$, the part due to the incident wave, a known function, and $\phi_d(x, y)$, the part due to the motion of the fluid because of the presence of the body (see Figure 5.3).

$$\phi = \phi_i + \phi_d \qquad (5.21)$$

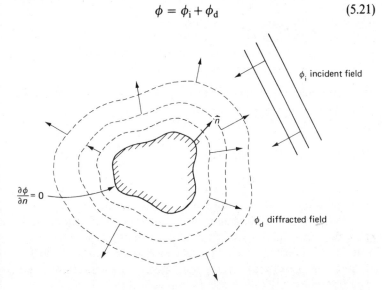

ϕ_i incident field

$\vec{n}$

$\frac{\partial \phi}{\partial n} = 0$

ϕ_d diffracted field

Figure 5.3   Incident and diffracted fields

Once the differential equation and boundary conditions are recast in terms of the unknown function ϕ_d these are solved and hence Φ is known. From Φ we may obtain the pressure of the fluid on the body from the Bernoulli equation, for an isentropic inviscid fluid:

$$\frac{\partial \Phi}{\partial t} + gz + \frac{1}{2} |\nabla \Phi|^2 + \frac{P}{\rho} = F(t) \tag{5.22}$$

where ρ is the density of the fluid and $F(t)$ is a function, constant in space, that we may set equal to zero. As our theory is linear we neglect the non-linear term $|\nabla \Phi|^2$. The force on the cylinder will then be obtained by integrating the force due to the pressure over the whole body, i.e.

$$\vec{F} = - \int_S P\vec{n} \, \mathrm{d}S \tag{5.23}$$

where $\vec{n}$ is a unit normal vector into the fluid.

To incorporate the effects of drag into this force we could evaluate an integral like $\displaystyle\int_S \sigma_{ij} n_i \, \mathrm{d}S$ over the surface of the cylinder,

where σ_{ij} is the rate of stress tensor

n_i is the unit normal vector to the surface

S is the intersection of a horizontal plane at depth z with the cylinder surface

This would include viscosity without taking into account the modification of the flow field by viscous effects.

The radiation condition

In the solution for ϕ (or ϕ_d) we usually have to resort to expressing ϕ_d as an infinite sum of eigenfunctions corresponding to the geometry with which we are dealing and satisfying Helmholtz's equation. We know that only the eigenfunctions corresponding to waves *diverging* from the solid obstacle are valid, and hence we have to have a method of choosing the eigenfunctions that satisfy this condition. Such a condition is usually called a 'radiation condition'.

Consider the wave equation for ϕ, now considered to be a function of the variables x, y and t, given by:

$$c^2 \nabla^2 \phi = \frac{\partial^2 \phi}{\partial t^2} \tag{5.24}$$

where c is the celerity of the wavelike disturbances that are solutions of this equation.

If we have to solve this equation in a bounded region S containing all the solid bodies and sources of the problem, we want a radiation condition, which can be imposed on the boundary Γ (in particular, in finite or boundary element methods, on the external element boundaries). See Figure 5.4.

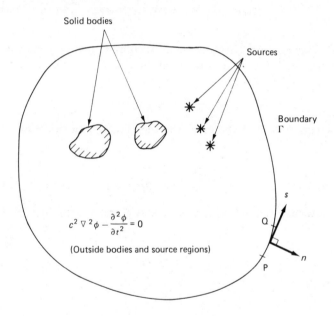

Figure 5.4   Wave equation problem region

Consider a *small* segment $\delta\Gamma$ of the boundary, and take a set of orthogonal local coordinates (s, n) on this segment, as shown in Figure 5.5. Near $\delta\Gamma$ the curve is locally straight, and in these coordinates we have:

$$c^2 \left(\frac{\partial^2 \phi}{\partial s^2} + \frac{\partial^2 \phi}{\partial n^2} \right) = \frac{\partial^2 \phi}{\partial t^2} \tag{5.25}$$

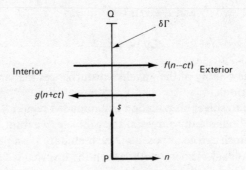

Figure 5.5 Small segment of boundary

As $\delta\Gamma$ is small we may assume that ϕ is approximately constant along the segment; hence we only consider the variation of ϕ in the normal direction n, and write:

$$\phi(s, n, t) = \phi_s \, \tilde{\phi}(n, t) \tag{5.26}$$

Now equation (5.25) becomes the one-dimensional wave equation for $\tilde{\phi}$:

$$\frac{\partial^2 \tilde{\phi}}{\partial n^2} = \frac{1}{c^2} \frac{\partial^2 \tilde{\phi}}{\partial t^2} \tag{5.27}$$

We now take characteristic coordinates ξ and η given by:

$$\xi = n - ct \tag{5.28}$$

$$\eta = n + ct$$

Then:

$$\frac{\partial}{\partial \xi} = \frac{\partial}{\partial n} - \frac{1}{c} \frac{\partial}{\partial t}$$

$$\frac{\partial}{\partial \eta} = \frac{\partial}{\partial n} + \frac{1}{c} \frac{\partial}{\partial t} \tag{5.29}$$

and (5.27) can be written:

$$\frac{\partial^2 \tilde{\phi}}{\partial \eta \, \partial \xi} = 0 \tag{5.30}$$

The general solution for this equation (first derived by D'Alembert) is:

$$\tilde{\phi}(\eta, \xi) = f(\xi) + g(\eta)$$

$$= f(n - ct) + g(n + ct) \qquad (5.31)$$

where f and g are any two functions with continuous first derivatives.

The function $f(n - ct)$ or $f(\xi)$ represents a wave of constant profile passing through the segment $\delta \Gamma$ in the direction of increasing n with a velocity c, i.e. a wave passing out of our region; $g(n + ct)$ represents a wave travelling into our region, and these are the solutions we wish to exclude. We therefore require for our radiation condition:

$$g(\eta) \equiv 0 \qquad \text{on } \delta\Gamma \qquad (5.32)$$

and hence:

$$\partial g / \partial \eta \equiv 0 \qquad \text{on } \delta\Gamma \qquad (5.33)$$

or more usefully:

$$\partial \tilde{\phi} / \partial \eta = 0 \qquad (5.34)$$

which in our local variables can be written, using the original potential ϕ, as:

$$\frac{\partial \phi}{\partial n} + \frac{1}{c} \frac{\partial \phi}{\partial t} = 0 \qquad (5.35)$$

This is the radiation condition to be applied on each small segment of the boundary Γ. (The function ϕ should also be bounded on Γ.)

If, in particular, we assume a harmonic time dependence for ϕ we may write:

$$\phi(x, y, t) = \phi(x, y) \exp(i\omega t) \qquad (5.36)$$

Then the wave equation (5.27) becomes:

$$\nabla^2 \phi + \kappa^2 \phi = 0 \qquad (5.37)$$

(the Helmholtz equation).

$$\kappa = \omega / c \qquad (5.38)$$

and the radiation condition (5.35) becomes:

$$\frac{\partial \phi}{\partial n} + i\kappa\phi = 0 \qquad (5.39)$$

Strictly speaking this boundary condition applies only if the boundary Γ is infinitely far from the solid bodies and sources producing the waves. An expression similar to (5.39) was deduced much more rigorously by Sommerfeld [4].

We know that our reduced potential ϕ satisfies the Helmholtz equation (5.37), and hence:

$$\nabla^2 \phi_d + \kappa^2 \phi_d = -(\nabla^2 \phi_i + \kappa^2 \phi_i) \tag{5.40}$$

but for an incident linear wave given by equation (5.16) we also have:

$$\nabla^2 \phi_i + \kappa^2 \phi_i = 0 \tag{5.41}$$

$$\therefore \qquad \nabla^2 \phi_d + \kappa^2 \phi_d = 0 \tag{5.42}$$

Hence we require:

$$\lim_{r \to \infty} \sqrt{r} \left(\frac{\partial \phi_d}{\partial r} + i\kappa\phi_d \right) = 0 \tag{5.43}$$

and

$$\lim_{r \to \infty} \sqrt{r} \, |\phi_d| < \infty \tag{5.44}$$

where r is a radial global coordinate and the $\sqrt{r}$ factor is derived from the *spreading factor*. This takes into account the energy spread in two dimensions from a wave source. As we learnt in Chapter 3, the energy of a wave is proportional to the square of the amplitude per unit length of wave crest and hence is proportional to ϕ_d^2 for the scattered field. As the same energy is distributed over larger and larger circles of circumference $2\pi r$ centred on a source, the amplitude of the waves must decrease like $1/\sqrt{r}$; this factor is the spreading factor.

We shall be dealing in particular with circular cylindrical obstacles and hence shall be using the eigenfunctions corresponding to a circle, which are in fact Bessel functions. Some properties of Bessel functions are given below. The interested reader should refer to standard works on Bessel functions, the most comprehensive of which is Watson [5].

Bessel functions

(a) Derivation of Bessel functions

Consider Helmholtz's equation in two dimensions:

$$\nabla^2 \phi + \kappa^2 \phi = 0 \tag{5.45}$$

In polar coordinates (r, θ) this becomes:

$$\frac{1}{r}\frac{\partial}{\partial r}\left(r\frac{\partial \phi}{\partial r}\right) + \frac{1}{r}\frac{\partial}{\partial \theta}\left(\frac{1}{r}\frac{\partial \phi}{\partial \theta}\right) + \kappa^2 \phi = 0 \tag{5.46}$$

assuming a separable solution:

$$\phi(r, \theta) = R(r)\,\Theta(\theta) \tag{5.47}$$

Substituting, we obtain, with a separation constant μ:

$$\frac{1}{\Theta}\frac{d^2\Theta}{d\theta^2} = -\mu^2 \Rightarrow \Theta = \exp(\pm i\mu\theta) \tag{5.48}$$

$$\frac{d^2 R}{dr^2} + \frac{1}{r}\frac{dR}{dr} + \left(\kappa^2 - \frac{\mu^2}{r^2}\right)R = 0 \tag{5.49}$$

This last is Bessel's equation, with solutions in terms of Bessel functions of r of order μ:

$$R(r) = J_\mu(\kappa r) \quad \text{Bessel function of the first kind, order } \mu$$

or $\quad R(r) = Y_\mu(\kappa r) \quad$ Bessel function of the second kind, order μ

We can also define:

$$H_\mu^{(1)}(\kappa r) = J_\mu(\kappa r) + iY_\mu(\kappa r) \tag{5.50}$$

a Hankel function of the first kind, and:

$$H_\mu^{(2)}(\kappa r) = J_\mu(\kappa r) - iY_\mu(\kappa r) \tag{5.51}$$

a Hankel function of the second kind. These two functions will also be solutions of Bessel's equation. Two typical solutions of Helmholtz's equation will be (for integral $\mu = n$):

$$\phi(r, \theta) = \sum_{n=-\infty}^{\infty} A_n J_n(\kappa r)\exp(+in\theta) \tag{5.52}$$

or $$\phi(r, \theta) = \sum_{n=-\infty}^{\infty} B_n H_n^{(2)}(\kappa r)\exp(+in\theta) \tag{5.53}$$

with A_n and B_n constants to be determined by the boundary conditions.

(b) Asymptotic forms of the Bessel functions

By looking at integral representation of the Bessel functions we may obtain the asymptotic behaviour by expanding the integrand and integrating the first few terms.

For large κr:

$$J_n(\kappa r) \sim \left(\frac{2}{\pi \kappa r} \right)^{\frac{1}{2}} \cos \left[\kappa r - \frac{\pi}{4} - \frac{n\pi}{2} \right] \qquad (5.54)$$

Thus J_n is analogous to the cosine function in one dimension, apart from a spreading factor, with the two-dimensional nature of the cylindrical solution:

$$Y_n(\kappa r) \sim \left(\frac{2}{\pi \kappa r} \right)^{\frac{1}{2}} \sin \left[\kappa r - \frac{\pi}{4} - \frac{n\pi}{2} \right] \qquad (5.55)$$

$$H_n^{(2)}(\kappa r) \sim \left(\frac{2}{\pi \kappa r} \right)^{\frac{1}{2}} \exp \left[-i \left(\kappa r - \frac{\pi}{4} - \frac{n\pi}{2} \right) \right] \qquad (5.56)$$

Thus:

$$H_n^{(2)}(\kappa r) \exp(i\omega t) \sim \left(\frac{2}{\pi \kappa r} \right)^{\frac{1}{2}} \exp[-i(\kappa r - \omega t)] \times$$

$$\times \exp \left[-i \left(\frac{\pi}{4} + \frac{n\pi}{2} \right) \right] \qquad (5.57)$$

which is an *outgoing* wave in two dimensions analogous to the solution $\exp[-i(\kappa x - \omega t)]$, which represents a wave travelling in the direction of increasing x in one dimension. Figure 5.6 shows graphs of $J_1(r)$ and $Y_0(r)$. It can be seen from these graphs that the functions look like sine waves whose amplitude decreases with increasing r. These are the sort of waves we would expect to be diffracted from a cylinder if we think of r as increasing distance from the axis.

These functions will be used to construct a solution of the problem of diffraction of waves by a vertical, stationary, circular cylinder, or a group of such cylinders.

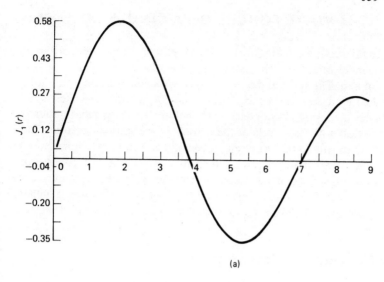

(a)

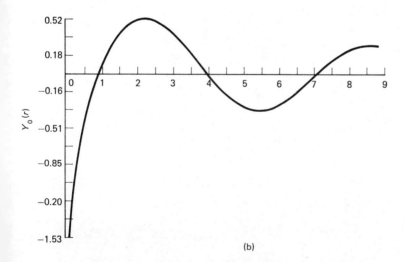

(b)

Figure 5.6   Graphs of (a) Bessel function of first kind, order 1, $J_1(r)$; (b) Bessel function of second kind, order 0, $Y_0(r)$

5.3 WAVE FORCES ON A SINGLE COLUMN

In this section we calculate the wave forces on a single vertical circular cylinder and, from this, find the force spectral density on a single column. The spectral density of the displacement of the column may then be found, and hence stress levels for certain sea-states may be calculated. Initially we consider the column to be rigid and stationary; we deal with the effects of the column's flexibility in section 5.7. An analysis similar to the one employed in this section may sometimes be used to calculate inertia coefficients for cylinders of other cross-sectional shapes, as mentioned in Chapter 4. We concentrate on the horizontal forces here, since the buoyancy effects may be calculated hydrostatically given the surface elevation on the cylinder surface.

The velocity potential

We consider a linear plane wave of amplitude a_0 and frequency ω impinging on a vertical circular cylinder, of radius a, that penetrates the water surface and reaches to the sea bed, which is flat, impermeable and at a depth d. The coordinates are taken as in Figure 5.7

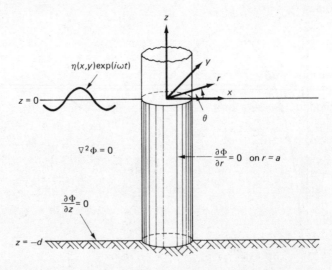

Figure 5.7   Coordinates and boundary conditions for a vertical circular cylinder

and the wave is travelling at α to the positive x axis. We assume potential flow and can write:

$$\nabla^2 \phi + \kappa^2 \phi = 0 \qquad (5.58)$$

and

$$\phi = \phi_i + \phi_d \qquad (5.59)$$

where

$$\phi_i = \frac{iga_0}{\omega} \exp[-i\kappa r \cos(\theta - \alpha)] \quad \text{(real part implied)} \quad (5.60)$$

the potential due to an incident linear wave travelling at an angle α to the positive x axis, which by Watson[5] as an expansion of the exponential in terms of Bessel functions gives:

$$\phi_i(r, \theta, \omega) = \frac{iga_0}{\omega} \sum_{-\infty}^{\infty} (-i)^n J_n(\kappa r) \exp[-in(\theta - \alpha)] \qquad (5.61)$$

But ϕ_i is a solution of Helmholtz's equation, and hence:

$$\nabla^2 \phi_d + \kappa^2 \phi_d = 0 \qquad (5.62)$$

and the boundary condition on the cylinder surface $(\partial\phi/\partial r = 0)$ becomes:

$$\frac{\partial \phi_i}{\partial r} = -\frac{\partial \phi_d}{\partial r} \qquad r = a \qquad (5.63)$$

So we must find a diffracted potential ϕ_d which satisfies Helmholtz's equation, represents waves radiating from the cylinder and satisfies the boundary condition (5.63). Such a ϕ_d can be written:

$$\phi_d(r, \theta, \omega) = \frac{iga_0}{\omega} \sum_{-\infty}^{\infty} \alpha_r H_n^{(2)}(\kappa r) \exp(-in\theta) \qquad (5.64)$$

where the unknown α_r $(r = 0, \pm 1, \pm 2 \ldots)$ are constants. These α_r are calculated by applying the boundary condition (5.63) and using the fact that the resulting expression is an identity, true for all values of θ, enabling us to equate coefficients of $\exp(-in\theta)$. We obtain:

$$\alpha_r = -(-i)^r \exp(+ir\alpha) \frac{J_r'(\kappa a)}{H_r^{(2)'}(\kappa a)} \qquad r = 0, \pm 1, \pm 2, \pm 3 \ldots$$

$$(5.65)$$

where ' denotes differentiation with respect to the argument of the function.

Substituting these values for α_r back in (5.64), we obtain an expression for the reduced potential throughout the fluid in the presence of the cylinder.

The force on the cylinder

By considering a small element of the surface of the cylinder the force on this element will be, in polar coordinates (r, θ) (see Figure 5.8):

$$\vec{F}(\omega) = (-P\vec{r}\,dS, 0)$$

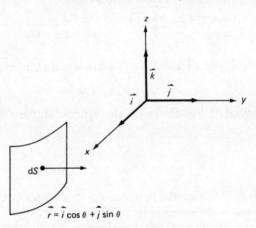

$$\vec{r} = \vec{i}\cos\theta + \vec{j}\sin\theta$$

Figure 5.8　Element on cylinder surface

or in Cartesian coordinates (x, y):

$$\vec{F}(\omega) = (-P\cos\theta\,dS, \; -P\sin\theta\,dS)$$

where P is the pressure of the fluid at the depth of the element, and $\vec{r}$ is the unit outward radial vector. So the total force on the cylinder will be:

$$\vec{F}_t(\omega, t) = -\int_S P\vec{r}\,dS = -\int_0^{2\pi}\int_{-d}^0 P\vec{r}a\,d\theta\,dz \qquad (5.66)$$

where S is the surface of the cylinder. The force per unit length at depth z is:

$$F(z, t, \omega) = - \int_0^{2\pi} P \vec{r} a \, d\theta \qquad (5.67)$$

From the linearised Bernoulli equation (5.22) we may obtain:

$$P = - \rho g z - \rho \frac{\partial \Phi}{\partial t} \qquad (5.68)$$

The first term represents the hydrostatic pressure and the second the dynamic pressure, in our linear approximation. Hence:

$$\vec{F}_t(\omega, t) = \int_{-d}^0 \int_0^{2\pi} \rho \frac{\partial \Phi}{\partial t} \vec{r} a \, dz \, d\theta + \int_{-d}^0 \int_0^{2\pi} \rho g z \vec{r} a \, d\theta \, dz \quad (5.69)$$

$$= 0 \text{ as } \int_0^{2\pi} \vec{r} \, d\theta = 0$$

the integration being taken only up to the still water level. Hence:

$$\vec{F}_t(\omega, t) = i\omega\rho a \int_{-d}^0 \frac{\cosh[\kappa(z+d)]}{\cosh \kappa d} \, dz \int_0^{2\pi} \phi(r, \theta) \Big|_{r=a} \vec{r} \, d\theta \exp(i\omega t) \qquad (5.70)$$

since $\partial\Phi/\partial t = i\omega\Phi$.

To perform the angular integration $\int_0^{2\pi} d\theta$, we need:

$$\int_0^{2\pi} \cos[n(\theta - \alpha)] \cos\theta \, d\theta = \pi \cos n\alpha \, \delta_{n1} \qquad (5.71)$$

and

$$\int_0^{2\pi} \cos[n(\theta - \alpha)] \sin\theta \, d\theta = \pi \sin n\alpha \, \delta_{n1} \qquad (5.72)$$

where

$$\delta_{n1} \begin{cases} = 0 & n \neq 1 \\ = 1 & n = 1 \end{cases}$$

Hence:

$$\vec{F}_t(\omega, t) = 4\rho g a_o \frac{\tanh \kappa d}{\kappa^2} \frac{\exp(i\omega t)}{H_1^{(2)'}(\kappa a)} (\cos \alpha, \sin \alpha) \qquad (5.73)$$

where the bracket denotes a two-dimensional Cartesian vector. The force is therefore in the direction of wave advance.

Taking the real part for the actual physical force, we obtain:

$$\vec{F}_t(\omega, t) = 4\rho g a_o \frac{\tanh \kappa d}{\kappa^2} \frac{[\cos \omega t \, J_1'(\kappa a) - \sin \omega t \, Y_1'(\kappa a)]}{[J_1'(\kappa a)]^2 + [Y_1'(\kappa a)]^2} (\cos \alpha, \sin \alpha)$$

$$(5.74)$$

This analysis was first carried out by MacCamy and Fuchs[6], although they are in error in choosing a Hankel function of the second kind in their expansions, since these have a time dependence $\exp(-i\omega t)$.

The peak value of the force $\vec{F}_{tpeak}(\omega)$ is given by the modulus of the complex expression (as the variation is harmonic), and equals $\sqrt{2} \times$ r.m.s. force:

$$\vec{F}_{tpeak}(\omega) = 4\rho g a_o \frac{\tanh \kappa d}{\kappa^2} \frac{1}{\left| H_1^{(2)'}(\kappa a) \right|} (\cos \alpha, \sin \alpha) \qquad (5.75)$$

The corresponding expressions for the force on unit length at depth z are:

$$\vec{F}(\omega, z, t) = \frac{4\rho g a_o}{\kappa} \frac{\cosh[\kappa(z+d)]}{\cosh \kappa d} \frac{\exp(i\omega t)}{H_1^{(2)'}(\kappa a)} (\cos \alpha, \sin \alpha)$$

$$(5.76)$$

the real part being:

$$\vec{F}(\omega, z, t) = \frac{4\rho g a_o}{\kappa} \frac{\cosh[\kappa(z+d)]}{\cosh \kappa d} \times$$

$$\times \frac{[\cos \omega t \, J_1'(\kappa a) - \sin \omega t \, Y_1'(\kappa a)]}{[J_1'(\kappa a)]^2 + [Y_1'(\kappa a)]^2} (\cos \alpha, \sin \alpha)$$

$$(5.77)$$

The peak value at depth z is then:

$$\vec{F}_{peak}(\omega, z, t) = \frac{4\rho g a_o}{\kappa} \frac{\cosh[\kappa(z+d)]}{\cosh \kappa d} \frac{1}{\left| H_1^{(2)'}(\kappa a) \right|} (\cos \alpha, \sin \alpha)$$

$$(5.78)$$

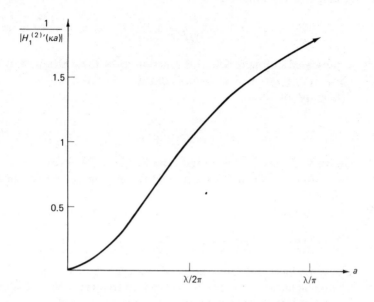

Figure 5.9   *Variation of inertia force with cylinder radius*

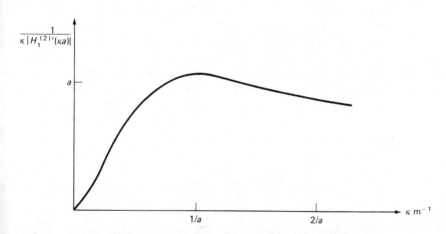

Figure 5.10   *Variation of inertia force with wavenumber of incident wave*

A graph of the function

$$\frac{1}{\left|H_1^{(2)'}(\kappa a)\right|}$$

is presented in Figure 5.9. This function gives some idea of how the above forces vary with cylinder diameter a. For deep water $\kappa d > 2.6$. The graph of

$$\frac{1}{\kappa\left|H_1^{(2)'}(\kappa a)\right|}$$

shown in Figure 5.10 is an indication of how the force varies with wavenumber κ (hence frequency ω) for a cylinder of a particular radius a.

Spectral analysis of forces

Above we have calculated the force on a cylinder under the influence of a plane linear wave of a particular fixed frequency. Since the main object of this chapter is to calculate the wave forces on a typical gravity offshore structure, some account must be taken of the random nature of the water waves found at that particular location. A spectral approach is used (see section 3.4).

The incoming (wind-generated) wave field is represented by a random superposition of harmonic periodic waves. For the present we assume that this wave field is unidirectional. The random incident surface elevation is assumed to be a zero-mean, Gaussian, ergodic process. A special feature of the linear approximation made is that the frequency of each spectral component determines the wavelength of that component uniquely; the amplitude of that component is determined by the use of some empirically derived expression, usually the spectrum of the wave heights $S_{\eta\eta}(\omega)$. The power spectrum of the wave forces $S_{FF}(\omega, z)$ may be given from equation (5.78) by:

$$S_{FF}(\omega, z) = \frac{16\rho^2 g^2}{\kappa^2} \frac{\cosh^2\left[\kappa(z+d)\right]}{\cosh^2 \kappa d} \frac{1}{\left|H_1^{(2)'}(\kappa a)\right|^2} S_{\eta\eta}(\omega) \quad (5.79)$$

The variance of the forces may be obtained by integrating the spectra over the frequency range, if this is desired.

We may compare these results with the corresponding results

obtained by using Morison's equation. Neglecting the drag term we obtain, from Chapter 4, the following force/unit length by Morison's equation:

$$S_{FF}^m(\omega, z) = 4\rho^2 a^4 \pi^2 g^2 \kappa^2 \frac{\cosh^2[\kappa(z+d)]}{\cosh^2 \kappa d} S_{\eta\eta}(\omega) \quad (5.80)$$

A comparison of these two sets of force spectral densities can be made *for any incident wave field spectral density* by considering the ratio:

$$\frac{S_{FF}^m(\omega, z)}{S_{FF}(\omega, z)} = \frac{S_{FFtot}^m(\omega)}{S_{FFtot}(\omega)} = \gamma(\omega, a, d) \quad \text{(say)} \quad (5.81)$$

$$\omega^2 = g\kappa \tanh \kappa d$$

where $$\gamma(\omega, a, d) = \frac{\pi^2}{4}(\kappa a)|H_1^{(2)'}(\kappa a)|^2 \quad (5.82)$$

In taking the ratio, the incident wave field spectral density cancels giving a measure of the proportional error of the Morison formula, which is a function of ω, a (the cylinder radius) and d (the water depth through κ), although the influence of the depth parameter is very small except for very long waves. We therefore deduce that, if $\gamma(\omega, a, d) < 1$ for some κa, we have a larger contribution from the corresponding frequency than would be expected from Morison's equation.

Figure 5.11 shows various plots of γ for different values of a throughout the range of frequencies of interest in offshore structure design. We see that for $a = 0.1$ m γ stays close to unity, whereas the curves for large radii of cylinders exhibit an initial dip followed by a

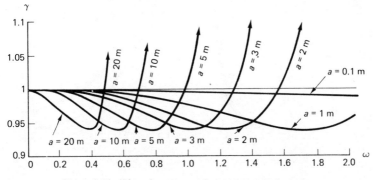

Figure 5.11 Plot of γ versus ω for varying cylinder radius

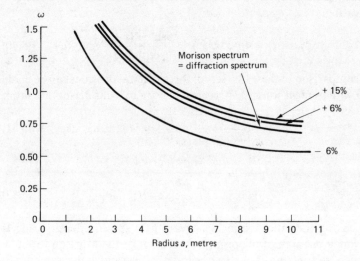

Figure 5.12   Percentage-error lines of the Morison force spectral density

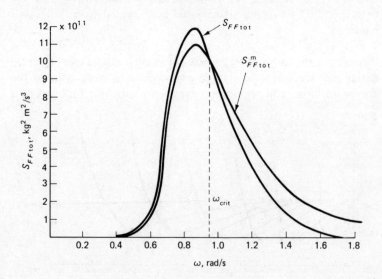

Figure 5.13   Wave force spectra

steep rise. These curves indicate that, below a certain frequency ω_{crit} (depending on a), the Morison force spectral density is an underestimate by as much as 6 per cent, whereas above $\omega = \omega_{crit}$ the Morison force becomes far too large, especially for large cylinders.

Figure 5.12 shows a plot of ω_{crit}; the region above this line is where the Morison equation overestimates the force, the region below is where the Morison equation underestimates the force by the percentage indicated on the other curves.

Figure 5.13 shows, for completeness, an actual plot of two total-force spectral densities for $a = 5$ m and $h = 100$ m, calculated using the Pierson–Moskowitz spectrum as the wave spectrum, for a wind speed of 10 m/s. One is calculated using the Morison formula, the other using diffraction theory. In this case it can be seen that the comparative displacement spectra of an offshore structure will depend on where the peak of the response function lies.

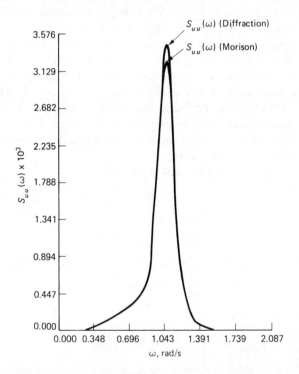

Figure 5.14   Morison and diffraction-theory displacement spectra
(cylinder radius $a = 2.6$ m)

Figure 5.14 shows two displacement spectral densities for a plane-frame offshore structure under the wave forces discussed so far. We see in this case, with cylinders of radius 2.6 m, that the spectral density calculated by diffraction theory is somewhat higher than the one calculated using Morison's equation.

Non-linear and viscous effects

There have been a number of attempts to use diffraction theory to take into account non-linearity in the waves, notably by Chakrabarti[7] and Oortmerssen[3]. Oortmerssen tried to take into account the constant resistance force experienced by a body in the presence of a wave train by using the non-linear (local kinetic energy) terms in the Bernoulli expression for the pressure. This analysis is dubious, however, because a *linear* wave theory is used and this effect is proportional to the square of the wave height. His results do, however, seem to match up with the measured results he quotes.

Chakrabarti, on the other hand, used Stokes's fifth-order theory for his incident waves, and deduced a force from diffraction analysis corresponding to each of the five component waves that go to make the non-linear wave of a fixed frequency (see Chapter 3). Having obtained these forces he recast the expressions in a form that corresponded to the Morison expression for inertial forces. In this way he derived five coefficients, each corresponding to one component of the non-linear wave.

The force per unit length was given by:

$$F(\omega, z, t) = \sum_{\alpha = 1}^{5} C_{m_\alpha} \rho \pi a^2 \dot{V}_\alpha \qquad (5.83)$$

where $\dot{V}_\alpha = \kappa \omega \alpha^2 \lambda_\alpha \cosh [\kappa(z + d)] \cos (\alpha \omega t - \delta_\alpha)$ (5.84)

λ_α is the usual amplitude parameter for fifth-order waves:

$$\tan \delta_\alpha = \frac{J_1'(\alpha \kappa a)}{Y_1'(\alpha \kappa a)} \qquad (5.85)$$

$$C_{m_\alpha} = \frac{4}{\pi (\alpha \kappa a)^2} \frac{1}{\{[J_1'(\alpha \kappa a)]^2 + [Y_1'(\alpha \kappa a)]^2\}^{\frac{1}{2}}} \qquad (5.86)$$

These force coefficients are shown as functions of the dimensionless parameter κa in Figure 5.15.

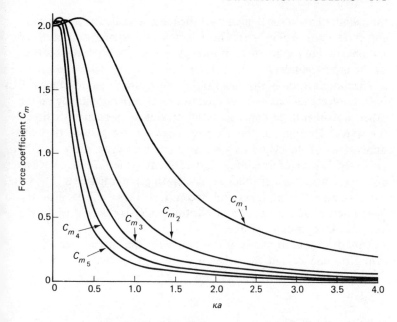

Figure 5.15   Values of effective inertia coefficients

In a later paper [8] Chakrabarti has attempted to solve the linearised Navier–Stokes equation in the presence of a vertical circular cylinder by decomposing the velocity field of the fluid into an irrotational part $V^{(i)}$ corresponding to potential theory and a rotational part $V^{(r)}$ that takes into account viscous effects. A Stokes fifth-order incident wave was used. In this way a Morison-type expression was deduced including a drag term. This term, however, is not directly comparable with the usual Morison drag term, as Chakrabarti points out. The Morison drag term is primarily a result of the formation of a wake behind the cylinder, whereas the drag term Chakrabarti derives is more like a tangential stress term. The analysis is very complex, so the interested reader is referred to the original paper for further details.

5.4 GROUPS OF CYLINDERS

It was mentioned in Chapter 3 that Morison's equation calculates the wave force on a cylinder without taking into account the effect that the cylinder has on the fluid motion; hence when calculating the wave forces on a group of cylinders no account is taken of the effect of the modification of the flow field by one cylinder on the incident flow field of the other cylinders.

The modification of the flow field by one cylinder on the other will include effects such as vortices shed by one cylinder impinging on the other, and other three-dimensional effects due to the small viscosity of the water. Throughout this chapter, however, we assume that the dimensions of the cylinders are such that these drag effects can be neglected. We therefore confine the treatment to inertia effects such as are experienced by a cylinder in accelerating potential flow.

Diffraction theory takes full account of these effects. For the mathematical details of the analysis the reader is referred to Spring and Monkmeyer[9] and Walker[10].

A modification to Morison's equation for the special case of two cylinders, one behind the other with respect to the wave field, is described at the end of this section.

Interaction between two cylinders

We now consider the forces on each of two circular, stationary cylinders standing in an incompressible fluid of depth h in the presence of a plane linear wave of angular frequency ω. See Figure 5.16.

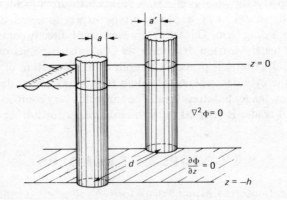

Figure 5.16 Two vertical cylinders with linear wave

We assess the effect the presence of one cylinder has on another by considering first the forces on one cylinder in isolation, and then the forces on the same cylinder in the presence of the other. The ratio of these forces will then be some measure of the interaction between the two cylinders.

If F_1^1 is the force on cylinder 1 in isolation and F_2^1 is the force on cylinder 1 in the presence of cylinder 2, the solid curve of Figure 5.17 is a plot of the ratio of these two forces F_2^1/F_1^1 for increasing separation of the two cylinders. The cylinder radius chosen here is 4 m, and the wavelength is 62.8 m; this corresponds to a wavenumber $\kappa = 1/10$.

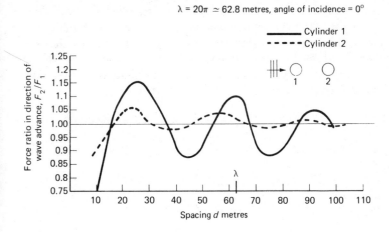

Figure 5.17   *Force ratio for two cylinders in line with direction of wave advance*

As can be seen from the figure, the interaction effect does *not* monotonically decrease with separation. The force on the first cylinder first increases, then decreases for increasing separation; this oscillatory behaviour is repeated for still increasing separation, another peak being reached for greater separation at about one wavelength. The second peak is, however, lower than the first, as would be expected from the two-dimensional nature of the problem.

Looking at the dotted curve in Figure 5.17 we see the same oscillatory behaviour. This curve is the ratio F_2^2/F_1^2 of the forces on the second cylinder alone and with the other cylinder present. We can see that the interaction effects are in fact smaller for this cylinder than for the leading one, so the term 'shielding' is inappropriate here.

Clearly there is a resonant effect in operation, and following Hogben[1] we may think of the region directly in front of the cylinders as a reflection region where the incident wave is almost totally reflected, as if from a plane wall. With this interpretation, we can see that if the reflected wave from the rear cylinder is in phase with the incident wave there will be constructive interference; if not, there will be some cancellation and a diminution of the force on the first cylinder. This explains the resonant-type behaviour: it is the relative phase of the two waves that is important.

To explain why the first cylinder experiences a larger interactive effect, consider the nature of our fluid flow approximations. With potential flow we have a symmetry of the streamlines for one cylinder about the axis through the cylinder parallel to the incoming wave crests; the streamlines therefore 'close up' after the cylinder, since there is assumed to be no boundary-layer separation. Thus the first cylinder would not be expected to affect the second so strongly, unless the two were very close.

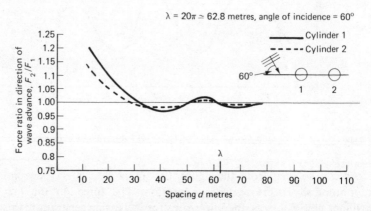

Figure 5.18   Force ratio for two cylinders at 60° to direction of wave advance

Figure 5.18 shows the force ratios when the two cylinders are inclined at an angle of 60° to the direction of wave advance. As expected, the interaction effects are smaller but still appreciable.

N cylinders (approximate solution)

The method referred to above for calculating forces involves the simultaneous solution of an infinite number of linear equations with Bessel-function coefficients. This is obviously very cumbersome and impractical, especially if we have more than two cylinders present. The analysis does, however, point to the simple approximation that may be made and that is, in many ways, reminiscent of the Morison equation.

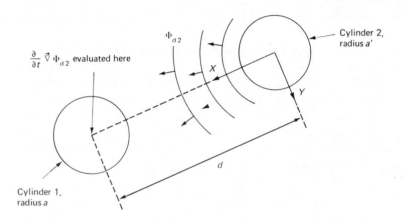

Figure 5.19 Interaction between two cylinders

If we have the situation depicted in Figure 5.19, following an analysis similar to that in references 9 and 10 we may write the force on cylinder 1 in the presence of cylinder 2 as:

$$F_x^1(z, t, \omega) = F_x + \frac{4i\omega\rho}{\kappa^2} \frac{1}{H_1^{(2)\prime}(\kappa a)} \frac{\partial}{\partial x} (\dot{\Phi}_{d2}) \bigg|_{X = d} \quad (5.87)$$

$$F_y^1(z, t, \omega) = F_y + \frac{4i\omega\rho}{\kappa^2} \frac{1}{H_1^{(2)\prime}(\kappa a)} \frac{\partial}{\partial y} (\dot{\Phi}_{d2}) \bigg|_{X = d} \quad (5.88)$$

where X and Y are measured from the centre of cylinder 2 (see Figure 5.19) and the differentials are evaluated at $X = d$, the separation of the cylinders. Here Φ_{d2} is the field diffracted from the second cylinder, and $\vec{F} = (F_x, F_y)$ is the force expected on the first cylinder in isolation. So, neglecting for the moment the effect that diffraction from cylinder 1 has on the field diffracted from cylinder 2, we may interpret the interaction terms as Morison-type terms that express the *extra force* in terms of the flow field at the axis of cylinder 1, i.e.

$$\vec{F}^1(z, t, \omega) = \vec{F} + \frac{4i\omega\rho}{\kappa^2} \frac{1}{H_1^{(2)\prime}(\kappa a)} \frac{\partial}{\partial t} (\vec{\nabla}\Phi_{d2}) \bigg|_{X=d} \qquad (5.89)$$

This result may then easily be extended to any number of cylinders to within the limits of the approximation.

5.5 SUBMERGED BODIES

In this section we look at the wave forces on submerged bodies, in particular vertical circular cylinders. The results will be particularly useful when we need to consider the forces on submerged oil storage tanks and the forces on the bases of offshore gravity production platforms. We start by looking at a diffraction approach to this problem, and continue by giving the results of various numerical studies of this and related problems.

Formulation of the problem

We consider a linear plane harmonic wave of amplitude a_o and frequency ω impinging on a submerged vertical circular cylinder, radius b, of height h_b in water of uniform depth d. Without loss of generality, by the symmetry of the problem we may assume the wave to be travelling in the positive x direction. We assume potential flow and hence the usual boundary conditions still hold. In addition to these conditions we have the no-flux boundary conditions (no flux through the cylinder surfaces) for the vertical curved surface and the horizontal top surface (see Figure 5.20). We now examine the situations in which we may relax the top surface boundary condition.

Consider the particle orbits within the fluid for a progressive linear wave. We have, for the velocity potential at depth z and time t:

$$\Phi(x, y, z, t) = \frac{ig}{\omega} a_o \exp[-i(\kappa x - \omega t)] \frac{\cosh[\kappa(z+d)]}{\cosh \kappa d} \qquad (5.90)$$

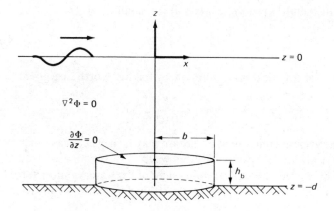

Figure 5.20 Submerged circular cylinder

Following Milne Thomson[11] (p. 431), we obtain the particle orbits as the ellipses:

$$\frac{x'^2}{\alpha^2} + \frac{z'^2}{\beta^2} = 1 \qquad (5.91)$$

where

$$\alpha = \frac{a_o \cosh\left[\kappa(z+d)\right]}{\cosh \kappa d}, \quad \beta = \frac{a_o \sinh\left[\kappa(z+d)\right]}{\cosh \kappa d}$$

and x' and z' are the horizontal and vertical distances, respectively, measured from some fixed reference point at depth z below the still water level, $z = 0$. Since

$$\alpha^2 - \beta^2 = \frac{a_o^2}{\sinh^2 \kappa d} \qquad (5.92)$$

all the ellipses have the same distance between their foci, and the lengths of their axes decrease as we go downwards into the liquid. At $z = -d$, $\beta = 0$, so the ellipse degenerates into a horizontal straight line and the particles move horizontally.

From this argument we can deduce that the vertical component of the particle motion may be neglected to within 5 per cent when $\beta/\alpha \leqslant 1/20$, i.e. at a depth z given by:

$$\tanh\left[\kappa(z+d)\right] \leqslant \frac{1}{20} \qquad (5.93)$$

or, since $\tanh\left[\kappa(z+d)\right] \simeq \kappa(z+d)$ for small $\kappa(z+d)$:

$$z+d \leqslant \frac{1}{20\kappa} \tag{5.94}$$

which for a typical wave component for the North Sea gives:

$$z+d \leqslant \frac{1}{0.02} = 50$$

so the vertical component of motion may be neglected as much as 50 metres above the sea bed for this wave.

If we neglect the vertical component of the water motion at the top of the cylinder, $z = -(d-h_b)$, to within 1 per cent we must have:

$$h_b \leqslant \frac{1}{100\kappa} \tag{5.95}$$

for the wave under consideration. We can then consider the fluid motion in the region $z > -(d-h_b)$ above the cylinder to be virtually unaffected by the presence of the cylinder, so our velocity potential will be just the incident velocity potential due to the incoming wave. In the region $z < -(d-h_b)$, by the same argument, we can say that the flow is approximately horizontal and so we treat this as a diffraction problem as before.

The force on the cylinder is calculated in exactly the same way as before, giving the force per unit length at depth z in the x direction:

$$\vec{F}^b(\omega, z, t) = \frac{4\rho g a_o}{\kappa} \frac{\cosh\left[\kappa(z+d)\right]}{\cosh \kappa d} \frac{\exp(+i\omega t)}{H_1^{(2)\prime}(\kappa b)} (1, 0) \tag{5.96}$$

The real part gives the physical force and the modulus the peak value. The total force on the cylinder is obtained by integrating this expression over the cylinder length, i.e.

$$\vec{F}_t^b(\omega, t) = \int_{-d}^{-(d-h_b)} \vec{F}^b(\omega, z, t)\, dz \tag{5.97}$$

$$\therefore \qquad \vec{F}_t^b(\omega, t) = \frac{4\rho g a_o}{\kappa^2} \frac{\sinh\left[\kappa h_b\right]}{\cosh\left[\kappa d\right]} \frac{\exp(+i\omega t)}{H_1^{(2)\prime}(\kappa b)} (1, 0) \tag{5.98}$$

The force per unit length for an immersed cylinder is the same as that for a complete cylinder, i.e. one that pierces the water surface. The total force, on the other hand, is the same as that for a complete cylinder, apart from a factor:

$$\frac{\sinh[\kappa h_b]}{\sinh[\kappa d]} = \frac{h_b}{d} \quad \text{for small } (\kappa d) \qquad (5.99)$$

Example 5.1

The dimensions used in this example are typical for a North Sea structure. —

See Figure 5.21; let $a = 5\,\text{m}$, $b = 30\,\text{m}$, $h_b = 50\,\text{m}$, $d = 150\,\text{m}$, and

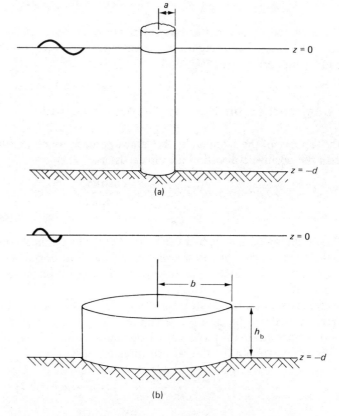

Figure 5.21 *Comparison between horizontal forces for base (b) and supporting column (a)*

$\kappa = 0.01094\,\text{m}^{-1}$ (corresponding to maximum wave height from Pierson–Moskowitz spectrum for wind speed of 30 m/s). Then:

$$\frac{F_a}{F_b} = \frac{\tanh\left[0.01094 \times 150\right]}{\left[H_1^{(1)\prime}(0.01094 \times 5)\right]} \times \frac{\cosh\left[0.01094 \times 150\right]}{\sinh\left[0.01094 \times 50\right]} \times$$

$$\times \left[H_1^{(1)\prime}(0.01094 \times 30)\right]$$

$$\simeq \frac{1}{8}$$

where F_a = total peak horizontal force on supporting column in direction of wave advance

$\quad\quad F_b$ = total peak horizontal force on base in direction of wave advance

So typically the force on the base of a single-cylinder offshore structure is approximately eight times greater than the force on the cylindrical support.

Vertical forces on the top face of a cylinder

On the top face of the cylinder, by the same approach we obtain the force in the positive z direction (i.e. upwards) as:

$$F_t^{\text{top}}(t, \omega) = -\left[\rho g(d - h_b)\pi b^2\right]\hat{\mathbf{z}} - 2\pi b\rho g a_o \frac{\cosh\kappa h_b}{\cosh\kappa d} J_1(\kappa b)\exp(i\omega t)\hat{\mathbf{z}}$$

$$(5.100)$$

The first term represents the hydrostatic part of the force, the second term the dynamic part due to the motion of the fluid horizontally across the top of the cylinder. This force depends critically on the ratio of b to λ, another resonant effect.

The force on an elemental strip parallel to the incident wave crests can be calculated in a similar way (see Figure 5.22). Considering a strip of width dx, parallel to the y axis and at a distance x from the centre of the cylinder, we obtain for the vertical force on this strip:

$$\vec{F}^{\text{top}}(x, t, \omega) = -\rho g\left\{a_o\left(\frac{2\cosh(\kappa h_b)}{\cosh\kappa d} + 1\right)\exp[-i(\kappa x - \omega t)] + \right.$$

$$\left. + (d - h_b)\right\}\sqrt{(b^2 - x^2)}\hat{\mathbf{z}}\,dz$$

$$(5.101)$$

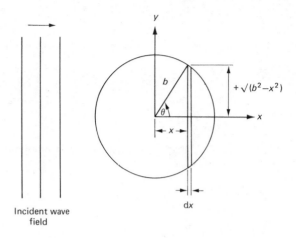

Figure 5.22   Plan view of submerged cylinder

This force demonstrates the existence of an overturning moment on the base. These forces will be transmitted directly to the underlying sea bed.

Other shapes

Most work on calculating wave forces on submerged bodies of other shapes has been numerical. The notable exception is that of Chakrabarti and Naftzger[12], who have calculated these forces for a submerged half-cylinder with a horizontal axis (Figure 5.23), and a hemisphere resting on the sea bed. The method used is similar to the one above, in that a decomposition into an incident field and a diffracted field is made, and each is expanded in terms of the eigenfunctions of the Laplace equation corresponding to the geometry of the obstacle considered. These two shapes were in fact considered to be in a non-linear, Stokes fifth-order wave field. The diffracted potential, as stated in reference 12, does not satisfy the free-surface boundary condition, so the results are best for a radius-to-depth ratio of less than 2 for the hemisphere and less than 4 for the half-cylinder.

For the half-cylinder the two force components are given by:

$$F_x = 2\rho V \dot{v}_x \qquad (5.102)$$

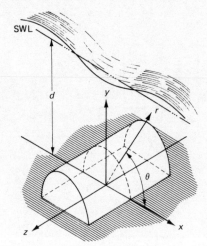

Figure 5.23   Submerged half-cylinder

where the displaced volume $V = \frac{1}{2}\pi a^2 l$, and:

$$F_y = \frac{2\rho\pi l g a_o \cos \omega t}{\cosh \kappa d} C_1(\kappa a) \quad \text{(oscillatory part only)} \quad (5.103)$$

where

$$C_1(\kappa a) = \cos \kappa a + \frac{\sin \kappa a}{\kappa a} + \kappa a S_i(\kappa a) - 1 \quad (5.104)$$

$S_i(\kappa a)$ is the sine integral, defined by:

$$S_i(\kappa a) = \int_0^{\kappa a} \frac{\sin x}{x} \, dx \quad (5.105)$$

The function $C_1(\kappa a)$ is plotted in Figure 5.24.

For the hemisphere the horizontal force is:

$$F_x = 1.5\rho V \dot{v}_x \quad (5.106)$$

where $V = 2\pi a^3/3$, and the oscillating part of the vertical force is:

$$F_y = \frac{\rho\pi g a_o \, a^2 \cos \omega t}{\cosh \kappa d} C_2(\kappa a) \quad (5.107)$$

where

$$C_2(\kappa a) = \sum_{n=0}^{\infty} \frac{(-1)^n}{n!(n+1)!} \frac{(1+4n)}{(1-4n^2)} \left(\frac{\kappa a}{2}\right)^{2n} \quad (5.108)$$

The function $C_2(\kappa a)$ is plotted in Figure 5.25.

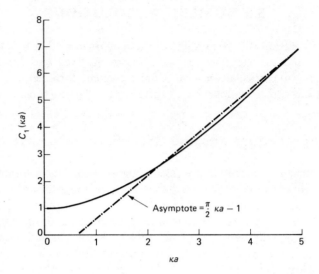

Figure 5.24   Function $C_1(\kappa a)$ associated with vertical force on a half-cylinder

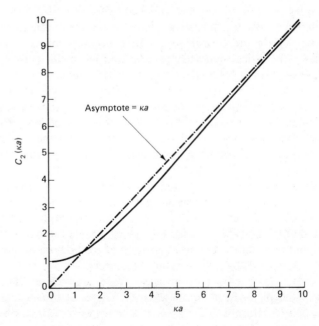

Figure 5.25   Function $C_2(\kappa a)$ associated with vertical force on a hemisphere

5.6 NUMERICAL SOLUTIONS

In this section we review some of the numerical methods that have been used to determine the wave forces on fixed and oscillating bodies. In particular we give a finite element formulation of the problem which is applicable to surface-piercing structures.

Integral equations and Green's function methods

The most general method for calculating wave forces was formulated by F. John[13] in 1950. This method is applicable to arbitrarily shaped bodies, and involves using the exact Green's function given in his paper and Green's theorem to obtain an integral equation for the potential over the boundary surfaces of the problem. Variations of this method involve using a Green's function (or fundamental solution) that does not satisfy all the boundary conditions, for example the free-surface boundary condition. The result of this is that various quantities then have to be integrated over the surfaces on which they do not fulfil the necessary conditions. Another important variation of the method is developed by Black[14]: he uses a symmetrical Green's function which is separable in the coordinates. This enables problems involving bodies with vertical symmetry to be solved more efficiently.

Published results using the above method abound in the literature, notably Boreel[15] and Hogben and Standing[2]. In these applications the solid body is represented by a matrix of sources whose magnitude is such as to cancel out the flux of fluid through the body surface. Figure 5.26 shows some results from reference 2 for the particularly interesting cases of submerged cylinders of circular and square cross-section.

The integral equation method can also be used to calculate the inertia forces on bodies oscillating with small amplitude. An exhaustive list of the literature on this subject is given in Wehausen[16]. The motion of the body is modelled by replacing the normal flux of fluid through the surface of the mean position of the body, resulting from the body motion, by a distribution of sources.

Another important tool in wave force analysis is Haskind's relations, explained fully in Newman[17]. These relations exploit the reciprocity between the radiation problem (of waves from a vibrating body) and the diffraction problem (of waves scattering from a fixed body). Haskind's theorem provides a method of evaluating the wave forces on a fixed body, from the radiated wave amplitude that would result from the same body oscillating.

Finite element methods

Finite element methods can often be applied to fluid flow problems as easily as they have been, more classically, to elasticity problems (Connor and Brebbia[18]). The refraction–diffraction problem has been tackled in this way by Berkhoff[19]. In his paper Berkhoff discretised the region surrounding an obstacle and regions of variable depth with a finite element mesh (see Chapter 8), and solved his wave propagation equation on this mesh for various circular islands and shoals. His method of applying the radiation condition, however, seems somewhat circuitous, and follows a similar treatment to that of Chen and Mei[20].

Here we present a finite element analysis of the diffraction problem, and for simplicity consider vertical structures in water of constant depth[21]. We assume that:

● The fluid is irrotational and incompressible.
● The motions are small.
● Far from the body the incident harmonic wave is undisturbed.

As is usual in diffraction analysis we split the velocity potential of the flow into an incident and a diffracted part, and we have the Helmholtz equation as the governing equation of our problem.

On a circle sufficiently far from the body (radius r) we can apply the following radiation condition (see Figure 5.27):

$$\frac{\partial \phi_D}{\partial n} + \frac{1}{c} \frac{\partial \phi_D}{\partial t} = 0 \qquad (5.109)$$

where c is the wave celerity ($c = \omega/\kappa$). This implies:

$$\frac{\partial \phi_D}{\partial n} = -i\kappa \phi_D \qquad (5.110)$$

In terms of the total potential we have:

$$\frac{\partial \phi}{\partial n} + i\kappa \phi = \frac{\partial \phi_I}{\partial n} + i\kappa \phi_I$$

$$(5.111)$$

or $\qquad \frac{\partial \phi}{\partial n} + i\kappa \phi = f \quad \text{(say)}$

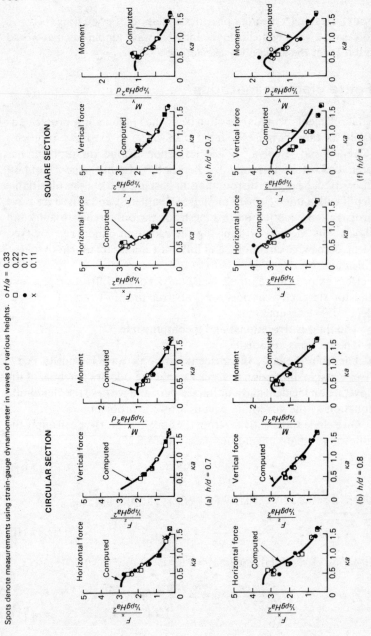

Spots denote measurements using strain-gauge dynamometer in waves of various heights.
o $H/a = 0.33$
□ 0.22
● 0.17
× 0.11

Figure 5.26 Comparison between computed and measured forces and moments on vertical columns (from Hogben and Standing[2])

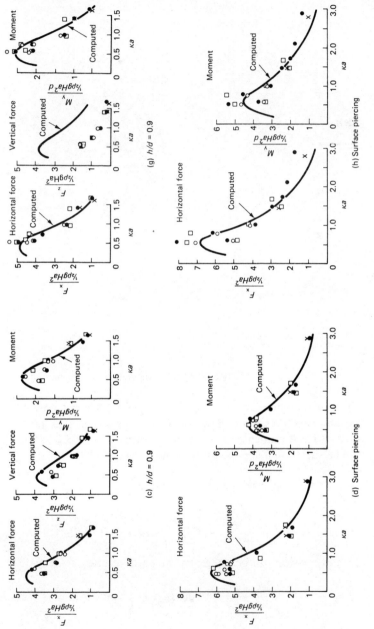

Figure 5.26 continued

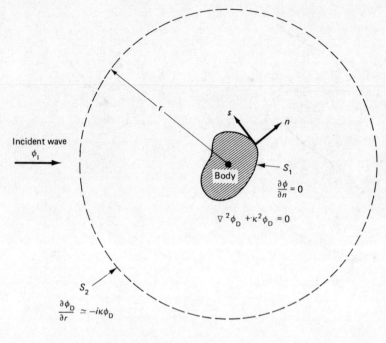

Note that: $\dfrac{\partial \phi}{\partial n} = \dfrac{\partial \phi_1}{\partial n} + \dfrac{\partial \phi_D}{\partial n}$

Figure 5.27 Diffraction problem

which is the boundary condition to use on S_2. We can now write a variational expression in terms of ϕ_D and use a numerical technique such as finite elements to solve the problem.

The variational expression in terms of ϕ is:

$$\iint\limits_{A} (\nabla^2 \phi + \kappa^2 \phi)\delta \phi \, \mathrm{d}A = \int_{S_1} \left(\frac{\partial \phi}{\partial n}\right)\delta \phi \, \mathrm{d}S +$$
$$+ \int_{S_2} \left(\frac{\partial \phi}{\partial n} + i\kappa \phi - f\right)\delta \phi \, \mathrm{d}S \tag{5.112}$$

This expression can be integrated by parts to give:

$$\iint \left(\frac{\partial \phi}{\partial x}\frac{\partial \delta \phi}{\partial x} + \frac{\partial \phi}{\partial y}\frac{\partial \delta \phi}{\partial y} - \kappa^2 \phi \, \delta \phi\right)\mathrm{d}A + \int_{S_2} i\kappa \phi \, \delta \phi \, \mathrm{d}S = \int_{S_2} f \delta \phi \, \mathrm{d}S \tag{5.113}$$

We now assume that the variable ϕ can be approximated on each element by:

$$\phi = \mathbf{G}^T \mathbf{\Phi}_e \qquad (5.114)$$

where $\mathbf{G}$ is an interpolation function vector and $\mathbf{\Phi}_e$ are the nodal unknowns. Typical interpolation functions are described in section 8.3.

Substituting (5.114) into (5.113), we obtain the following expression for an element:

$$\delta\mathbf{\Phi}_e^T \left\{ \int\!\!\int \left(\frac{\partial\mathbf{G}}{\partial x}\frac{\partial\mathbf{G}^T}{\partial x} + \frac{\partial\mathbf{G}}{\partial y}\frac{\partial\mathbf{G}^T}{\partial y} - \kappa^2 \mathbf{G}\mathbf{G}^T \right) dA + \right.$$

$$\left. + i\int_{S_2} \kappa\mathbf{G}\mathbf{G}^T dS \right\} \mathbf{\Phi}_e = \delta\mathbf{\Phi}_e^T \left(\int_{S_2} \mathbf{G}f\, dS \right) \qquad (5.115)$$

This expression can be written as:

$$\delta\mathbf{\Phi}_e^T (\mathbf{K} - \kappa^2 \mathbf{M} + i\kappa\mathbf{M}')\mathbf{\Phi}_e = \delta\mathbf{\Phi}_e \mathbf{F} \qquad (5.116)$$

where $\mathbf{K} = \int\!\!\int \left(\dfrac{\partial\mathbf{G}}{\partial x}\dfrac{\partial\mathbf{G}^T}{\partial x} + \dfrac{\partial\mathbf{G}}{\partial y}\dfrac{\partial\mathbf{G}^T}{\partial y} \right) dA$

$\mathbf{M} = \int\!\!\int \mathbf{G}\mathbf{G}^T dA$

$\mathbf{M}' = \displaystyle\int_{S_2} \mathbf{G}\mathbf{G}^T dS$

$\mathbf{F} = \displaystyle\int_{S_2} \mathbf{G}f\, dS$

We can now assemble the different matrices defined by the above equations into the global matrices for the whole continuum. This gives

$$\{ \mathscr{K} - \kappa^2 \mathscr{M} + i\kappa\mathscr{M}' \}\mathbf{\Phi} = \mathscr{F} \qquad (5.117)$$

(For further details regarding finite elements see Chapter 8.) This set of linear equations may then be solved on a computer and we may obtain the values of the velocity potential at each point of the mesh. These values having been obtained, the force on the obstacle may be obtained in the usual way by (numerical) integration around the surface of the obstacle represented by the contour S_1.

Example 5.2 (Diffraction by a cylinder)

The case of a single vertical column subject to an incident harmonic wave was studied.

The column was surrounded by a finite element mesh as shown in Figure 5.28, and wave diffraction results were found for different wavelengths, with the application of condition (5.110) on the external boundary.

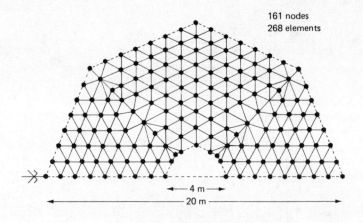

161 nodes
268 elements

4 m
20 m

Figure 5.28   Finite element mesh

As a test to determine the adequacy of the radiation condition in representing a train of plane harmonic waves, the case with no solid cylinder was first studied. For long waves (wavelength $\lambda = 30$ m) the results were very accurate, within 3 per cent of the exact solution. When the wavelength was reduced the errors tended to increase, which was to be expected as the element mesh became too coarse. It was found that, for linear elements such as the ones used here, approximately eight elements per wavelength should be used.

A test was run to compare the authors' results with those obtained by Chen and Mei[20], who used a finite element mesh and the Hankel function formulation, i.e. the fundamental solution. The results are for an incident wave with wavelength $\lambda = 2\pi$ and unit incident surface elevation for frequency $\omega = 3.1321$. Figure 5.29 compares the results of the exact solution, the finite element solution due to Chen and Mei, and the authors' solution[23]. The last compares favourably, taking into

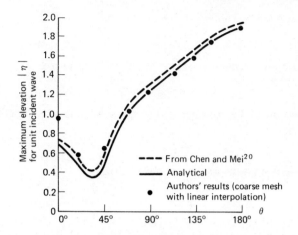

Figure 5.29 *Results for maximum surface elevations around a circular cylinder; wavelength = 2π metres, cylinder radius = 2 metres*

consideration the coarseness of the mesh around the cylinder. That of Chen and Mei, for instance, uses 18 elements round the cylinder and represents better the geometry of the obstruction.

5.7 EFFECTS OF COLUMN MOTION

A supporting column in the sea will be set in motion by wave forces to some extent, and in fact generates waves which will propagate outwards from the column. This wave radiation will carry away momentum and there will be a back reaction on the column. For a general treatment of this problem the numerical methods mentioned in the previous section should be used.

The importance of these radiated waves, in our calculations, will depend on the diameter of the vibrating member and the amplitude of its vibration. Although wider members will produce more waves they are less likely to vibrate with any appreciable amplitude. The radiation effect will, however, be important for large-diameter members when the structure is being bodily shaken horizontally, as it would be in an earthquake. The case of a translating cylinder is dealt with in this section.

If a structure is oscillating with a given frequency ω we have, for deep water ($\kappa d > 2.6$):

$$\omega^2 \simeq g\kappa \qquad (5.118)$$

and we obtain radiated waves of wavelength λ, where:

$$\lambda = 2\pi g/\omega^2 \qquad (5.119)$$

Hence we need to consider only waves of this wavelength and frequency.

Radiation of waves in an earthquake

During an earthquake an offshore structure may be considered to be moved horizontally, and if the members are sufficiently thick and rigid this motion may be thought of as a rigid-body translation. In keeping with the spectral approach of this book, we may consider one harmonic component of this motion.

In this section we consider particularly the motion of a surface-piercing vertical circular cylinder. Cases of oscillating submerged and floating cylinders have been dealt with extensively in the literature by numerical and variational techniques; the reader is referred particularly to Black, Mei and Bray[22].

In the present context we may obtain a simple analytical solution for the force on an oscillating surface-piercing vertical cylinder. With the usual notation we then have a modal shape given by:

$$f(z) \equiv X_m V z \qquad (5.120)$$

A Green's function method may be used to obtain the far field for the radiated waves, and *in this case* the momentum integral theorem may be used to deduce directly the reaction on the cylinder due to its motion. The total reaction is given by:

$$F(\omega, t) = 4\rho g X_m \frac{\tanh \kappa d}{\kappa^2} \frac{\exp(i\omega t)}{H_1^{(2)'}(\kappa a)} \qquad (5.121)$$

This force is in a direction opposite to the direction of oscillation.

The force per unit length at depth z is:

$$F(\omega, z, t) = \frac{4\rho g X_m}{\kappa} \frac{\cosh\left[\kappa(z+d)\right]}{\cosh \kappa d} \frac{\exp(i\omega t)}{H_1^{(2)'}(\kappa a)} \qquad (5.122)$$

The similarity of this expression to that obtained for the corresponding diffraction force is clear. The variation of this force with the cylinder diameter is of a similar nature to the corresponding variation of the diffraction force. This variation is represented in Figures 5.9 and 5.10. The frequency ω now represents the frequency of oscillation of the cylinder.

References

1. Hogben, N. G., *Fluid loading of offshore structures, a state of art appraisal: Wave loads*, R. Inst. Naval Arch. (1974)
2. Hogben, N. G., and Standing, R. G., 'Experience in computing wave loads on large bodies', OTC 2189, Proc. Offshore Technology Conf., Houston (1975)
3. Oortmerssen, G., 'The interaction between a vertical cylinder and regular waves', Symp. on Offshore Hydrodynamics, Wagenigen (1971)
4. Sommerfeld, A., *Partial differential equations in physics*, 256 and 344, Academic Press, New York (1949)
5. Watson, G. N., *A treatise on the theory of Bessel functions*, 2nd edn, Cambridge U.P. (1958)
6. MacCamy, R. C., and Fuchs, R. A., *Wave forces on piles: a diffraction theory*, US Army Coastal Engineering Center, Tech. Mem. No. 69 (July 1945)
7. Chakrabarti, S. K., 'Nonlinear wave forces on vertical cylinder', *Proc. ASCE (Hydraulics Div.)*, **98**, HY11, 1895–1909 (Nov. 1972)
8. Chakrabarti, S. K., 'Wave forces on piles, including diffraction and viscous effects', *Proc. ASCE (Hydraulics Div.)*, **99**, HY8, 1219–1233 (Aug. 1973)
9. Spring, B. H., and Monkmeyer, P. L., 'Interaction of plane waves with vertical cylinders', Proc. Conf. on Coastal Engineering, Vol. 2, Chapter 107 (1974)
10. Walker, S., *Determination of the wave forces on offshore gravity platforms, by diffraction theory and spectral analysis*, Univ. Southampton Dept Civil Engineering, Tech. Report, CH 1/77
11. Walker, S., *The inertial interaction of the surface piercing columns of an offshore gravity platform*, to be published
12. Chakrabarti, S. K., and Naftzger, R. A., 'Nonlinear wave forces on half-cylinder and hemisphere', *Proc. ASCE (Waterways, Harbors and Coastal Engng Div.)*, **100**, WW3 (Aug. 1974)
13. John, F., 'On the motion of floating bodies II', *Commun. in Pure and Applied Maths*, **3**, 45 (1950)
14. Black, J. L., 'Wave forces on vertical axisymmetric bodies', *J. Fluid Mech.*, **67**, 369 (1975)
15. Boreel, L. J., 'Wave action on large offshore structures', Proc. Conf. at Instn Civil Engineers, London (1974)
16. Wehausen, J. V., 'The motion of floating bodies', *A. Rev. Fluid Mech.*, **3** (1971)
17. Newman, J. N., 'The exciting forces on fixed bodies in waves', *J. Ship Res.*, **6**, 10 (1962)
18. Connor, J. J., and Brebbia, C. A., *Finite element techniques for fluid flow*, Newnes–Butterworths (1976)
19. Berkhoff, J. C. W., 'Computation of combined refraction–diffraction', Conf. on Coastal Engineering, Instn Civ. Engrs (1973)
20. Chen, H. S., and Mei, C. C., 'Oscillations and wave forces in a man-made harbor in the open sea', 10th Symp. on Naval Hydrodynamics.
21. Brebbia, C. A., Walker, S., and Kavanagh, M., 'Wave oscillation problems in

deep and shallow water', Proc. Int. Conf. on Offshore Structures, Rio de Janeiro, VFRJ September 1977, Pentech Press

22. Black, J. L., Mei, C. C., and Bray, M. C. G., 'Radiation and scattering of water waves by rigid bodies', *J. Fluid Mech.*, **46**, 151–164 (1971)

23. Brebbia, C. A., and Walker, S., 'Simplified boundary elements for radiation problems' (research note), *Applied Mathematical Modelling*, **2**, 2 (June 1978)

6 Effects of Currents and Winds

6.1 INTRODUCTION

In section 4.1 we reviewed briefly the environmental forces that we would expect to act on a typical offshore structure; this chapter deals more fully with the loads due to currents and winds. First we shall present a short explanation of how currents are produced, particularly pure drift currents, which are essentially a result of wind stress acting on the surface of the sea. Fuller descriptions of this topic may be found in books on physical oceanography, notably Ekman[1], Defant[2] and Muir Wood[3].

The strongest and most persistent currents have their origin in tidal forces caused by the influence of the sun's and moon's gravitational fields on the oceans. However, the travelling tidal waves and their attendant currents (which are in the direction of the wave at the crest and in the opposite direction in the trough) are modified by Coriolis forces, wind forces and gradient forces, which are fictitious forces caused by the slope of the sea bed. Coriolis forces are also fictitious forces, and seem to act on any particle on the earth; they arise because the coordinate axes in relation to which we do our dynamic calculations are rotating in space, being fixed with respect to the earth. The effect of Coriolis force is to deflect currents to the right in the northern hemisphere and to the left in the southern hemisphere. In an idealised circular sea in the northern hemisphere, this force will cause the location of high and low water to rotate anti-clockwise around its

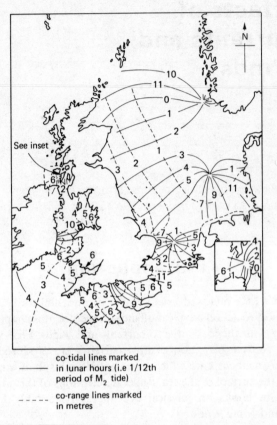

co-tidal lines marked
─────── in lunar hours (i.e 1/12th
period of M₂ tide)

- - - - - co-range lines marked
in metres

Figure 6.1   Tidal range and phase around British Isles, indicating amphidromal systems in the North Sea (reproduced from Admiralty Chart No. 5058 with the sanction of the Controller, HMSO, and the Hydrographer of the Navy)

perimeter, about an amphidromal point where there is no (semi-diurnal) tidal amplitude. In Figure 6.1 for the North sea we see that such points do occur in nature. On this map there are *co-tidal lines*, which are lines joining points with the same tidal phase, and *co-range lines*, which are lines joining points having the same range or tidal amplitude.

The atmosphere has three effects on the currents in the ocean. Firstly, and perhaps most simply, a region of high pressure in the atmosphere will tend to set up a barometric slope by causing a lowering of the sea surface by about one centimetre per millibar.

Secondly, the winds themselves may (as well as producing waves, see Chapter 3) produce a wind slope of the sea surface; for the relatively shallow waters of interest to the offshore engineer, this will be much more important than the barometric slope mentioned above. The third effect is that the wind stress acting on the surface of the sea will produce a pure drift current, which surprisingly moves with a velocity at the surface inclined at 45° to the direction of the wind. At greater depths this angle increases while the actual velocity of the current decreases. At the so-called *frictional depth* the current velocity will be in a direction opposite to the wind, but only of magnitude $\exp(-\pi)$ $\simeq 1/23$ of its surface value. Figure 6.2 shows the so-called Ekman Spiral of the velocity vectors of a pure drift current at varying depths. The direction and velocity of a current may therefore vary with depth, and even drift currents will not be in the same direction as the wind.

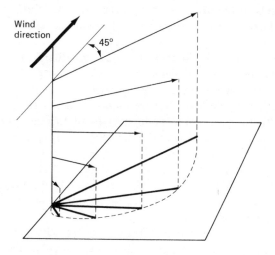

Figure 6.2   Vertical structure of pure drift current

In sections 6.2 to 6.5 we examine the effects that currents have on both large and slender members of an offshore structure, as well as the effects a current may have on the sea bed surrounding the structure's base—the very important scouring problem.

Sections 6.6 and 6.7 deal with winds, their characteristics and effect on offshore structures. As mentioned in Chapter 4, the wind forces represent typically only 5–10 per cent of the total environmental forces, acting only on the superstructure. During conditions of tow-out of a concrete or steel structure, however, these forces will become more important and must be allowed for when considering the stability of a structure in this situation.

6.2 CURRENTS

We have seen that currents will often have a velocity profile that decays very slowly with depth; this is particularly true of a pure tidal current arising from the propagation of very long tidal waves. In such a wave, as we saw in Chapter 3, the water particle motion is nearly horizontal and the decay with depth is given by a factor like $\exp(\kappa z)$, where the wavenumber $\kappa = 2\pi/\lambda$ and, as usual, z is negative. For long waves κ is very small and the decay with depth is very slow. So the current can be expected to have an influence over the whole immersed length of the structure, in contrast with the (wind-generated) waves of shorter wavelength.

The presence of a current has four main effects that we need to take into account in our force calculations.

1. The current will affect the water particle velocities of the surface waves; so, because the drag force on a member is proportional to the square of these velocities, a moderately small current may have a significant effect, particularly at large depths. This effect is very important for slender members, since the Morison equation used to calculate the forces on such members (equation 4.9) is modified both through the velocities and the coefficient C_D. The inertia force is not modified in this way, as it is assumed proportional to the water particle accelerations.

2. The current will also modify the surface wave field in a number of ways. The amplitude of the surface waves may be changed and some wave steepening may occur, as can be seen in the paper by Longuet–Higgins and Stewart[4]. The velocity of propagation and wavelength of the waves corresponding to a specific frequency will be altered also; in fact the waves can even be stopped by a current, which may cause a potentially dangerous concentration of wave energy (see Taylor[5]). Waves travelling obliquely over a current will also be refracted (Muir Wood[3], p. 65), and may be channelled along the current if the two directions form an angle greater than 90°. These effects will be small in the case of a North Sea design wave, however,

since the current speed is usually only 1.5 m/s whereas a typical wave speed will be of the order of 20 m/s.

3. A current impinging on any fixed body will cause the body to produce a standing wave pattern on the water surface and experience a corresponding reaction in the direction of the current. These waves are analogous to the 'ship waves' formed by a body moving through still water, with an obvious change of coordinate system (see Stoker[6] and Figure 6.3). Green's functions for such situations are known, and can be found in Wehausen and Laitone[7] and Lunde[8]. The methods described in section 5.6 may be used to solve such problems. This Green's function is very complicated, and the effort required for the calculation would not be justified except perhaps for large-diameter members in an estuary. Even in this situation good results are not expected, since flow separation and wake formation occur, making the assumptions of potential flow theory used in the calculations invalid.

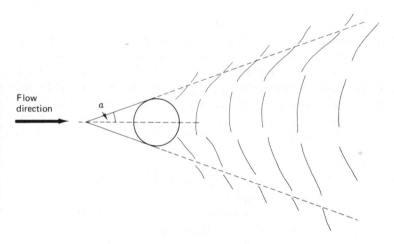

Figure 6.3   Standing wave pattern produced by stationary circular cylinder in uniform current

4. The fourth effect of a current was mentioned in Chapter 4 and is a very important effect for slender members, namely vortex shedding. See Figure 6.4. As a vortex detaches itself from the member it leaves behind an equal and opposite circulation around the member, which gives rise to a lift force perpendicular to the direction of the current.

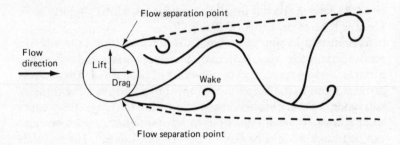

Figure 6.4 *Vortex shedding*

These vortices are often shed alternately from opposite sides of the member with a frequency related to the fluid flow velocity, Strouhal number and cylinder diameter. In this way there may be an oscillating force, transverse to the flow, that may coincide with the resonant frequency of that member, a potentially dangerous situation for steel lattice-type structures. This topic will be dealt with fully in section 6.4.

In addition we must bear in mind a number of factors that will determine the effectiveness of a current as an environmental load on a particular marine structure. This effectiveness will be governed by:

● The surface roughness of the member, possibly owing to the accretion of marine organisms during the life of the structure.

● The flexibility of the member, which may cause coupling to occur between the vortex shedding and the vibrational frequency of the member.

● The influence of nearby members; this may cause a resonant reaction, usually called 'galloping', which is a result of the wake of one member impinging on another.

● The dimensions of the member. Apart from affecting the Reynolds number of the flow around a member, the ratio of the diameter to the length of the member will affect the relative importance of the water surface effects, the vertical structure of the current and the effect that the current has on the conventional drag coefficient for that member.

Generally, the surface effects will be more important for large-diameter members such as the supporting columns of a concrete gravity-type platform, whereas effects such as the modification of the drag forces and galloping will be more important for slender members. The next section deals with the surface effects.

6.3 LARGE-DIAMETER MEMBERS

In this section we examine the surface effects of a current, which are of greater importance for large-diameter members. At a location where currents of varying direction are to be expected (i.e. where tidal currents dominate), the only reasonable shape for these members is the circular cylinder, which presents the same profile to the flow from all directions. We shall therefore restrict ourselves to this case.

Wave resistance

The waves produced by a current on a stationary body were mentioned in section 6.2 and give rise to a standing wave pattern. The angle α of Figure 6.3 does not depend on the current velocity, and is given by:

$$\alpha = \sin^{-1} \frac{1}{3} \qquad (6.1)$$

the disturbance being greatest along these lines. Waves of all frequencies and wavenumbers are generated by the body, but only reinforce in certain regions to reveal themselves in a pattern as shown in the figure. In the analysis this is represented by a stationary phase evaluation of an integral over all (vector) wavenumbers. Hogben and Standing[9] presented experimental results for this wave-making force, and defined a wave-making force coefficient C_W given by:

$$C_W = \frac{\text{wave resistance}}{\frac{1}{2}\rho A V_c^2} \qquad (6.2)$$

where A = projected frontal area
V_c = current speed (assumed constant)

Figure 6.5 shows experimental results for a circular cylinder. Hogben and Standing point out that, even to get approximate correspondence with results, an effectively elliptical section with

$$\varepsilon = b/a = 0.2$$

has to be used to model the flow separation and wake formation in the theoretical study. Here b is the radius of cylinder and a is the semi-axis of the ellipse perpendicular to the flow.

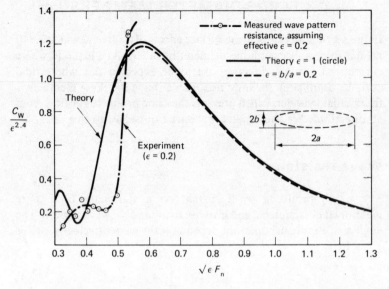

Figure 6.5   Wavemaking force on vertical circular cylinder, radius b
(from Hogben and Standing[9])

The ratio $C_W/\varepsilon^{2.4}$ is theoretically independent of ε when plotted against $\sqrt{\varepsilon}F_n$, for an elliptic cylinder with $\varepsilon = 0.2$, where F_n is the Froude number given by:

$$F_n = V_c/\sqrt{(2gb)} \qquad (6.3)$$

The theoretical results were derived by Kotik and Morgan[10]. In Hogben[11] the wave-making effect is discussed in the context of offshore structures in the North Sea, and it is shown that the wave-making forces are negligible in comparison with the drag forces, even in the fastest currents that could be expected.

Wave modification

It has been shown by Longuet–Higgins and Stewart[12] that the amplitude of short (wind-generated) waves is modified by any underlying long wave or tidal current. The current will not only convect the waves but will do work on the waves; Longuet–Higgins and Stewart termed this energy transfer the radiation stress. Figure

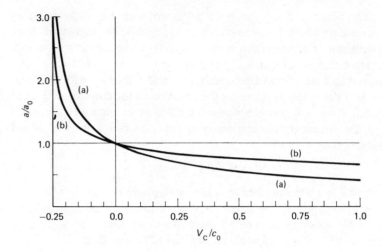

Figure 6.6   Amplification factor a/a_0 for waves on current of velocity V_c in direction of wave propagation: (a) with vertical upwelling from below; (b) with horizontal inflow from sides (from Longuet-Higgins and Stewart[12], courtesy Cambridge University Press)

6.6 shows the amplification factor a/a_0 for waves on a current V_c in the direction of wave advance. Here:

a = amplitude of wave on current of velocity V_c
a_0 = amplitude of same wave on stationary sea
c = wave velocity on current
c_0 = velocity of same wave on stationary sea

Curve (a) in the figure is plotted from:

$$\frac{a}{a_0} = \frac{c_0}{\sqrt{[c(c + 2V_c)]}} \qquad (6.4)$$

and is for a current with vertical upwelling to compensate for any mass imbalance due to current velocity change (a good model for a tidal current). Curve (b) in the figure is plotted from:

$$\frac{a}{a_0} = \frac{c_0}{\sqrt{\left[\dfrac{1}{c}(c + 2V_c)\right]}} \qquad (6.5)$$

which is for a current whose horizontal variation in V_c is compensated by a horizontal inflow from the sides.

The velocity c of waves on a current and the corresponding wavelength λ and wavenumber κ are given by kinematic considerations. For simplicity we shall consider waves on deep water in a current that is uniform and has magnitude V_c along the x axis. We now take a set of coordinate axes x', y' and z' moving with the fluid, and set the x' axis again along the direction of the current; see Figure 6.7(a). If we now consider waves travelling at an angle α to the x axis on the surface (in the stationary system), we can write the surface elevation as:

$$\exp[-i(\vec{\kappa} \cdot \vec{x} - \omega t)]$$ (6.6)

where $\vec{\kappa}$ is a wavenumber vector, of magnitude $\kappa = 2\pi/\lambda$ (where λ is the wavelength), in the direction of wave advance; see Figure 6.7(b).

In the moving-coordinate system this same wave field is given by $\exp[-i(\vec{\kappa}' \cdot \vec{x}' - \omega' t)]$, with an obvious notation. But:

$$\vec{x}' = \vec{x} - \vec{V}_c t$$ (6.7)

$$\therefore \exp[-i(\vec{\kappa}' \cdot \vec{x}' - \omega' t) = \exp\{-i[\vec{\kappa}' \cdot \vec{x} - (\omega' + \vec{V}_c \cdot \vec{\kappa}')t]\}$$

$$= \exp[-i(\vec{\kappa} \cdot \vec{x} - \omega t)]$$ (6.8)

Hence: $$\kappa \cos \alpha = \kappa' \cos \alpha'$$ (6.9)

where α' is the angle the waves appear to travel in the moving-coordinate system—Figure 6.7(c)

$$\omega' = \omega - \vec{V}_c \cdot \vec{\kappa} = \omega - V_c \kappa \cos \alpha$$ (6.10)

and the wave celerities are linked by:

$$c = c' + V_c$$ (6.11)

which gives: $$\frac{\cos \alpha}{\lambda} = \frac{\cos \alpha'}{\lambda'}$$ (6.12)

$$c' \sin \alpha' = c \sin \alpha$$ (6.13)

$$c' \cos \alpha' = c \cos \alpha - V_c$$ (6.14)

$$\omega' = \omega - \frac{V_c}{\lambda} \cos \alpha$$ (6.15)

The dispersion relation in the moving-coordinate system $\omega'^2 = g\kappa'$ becomes:

$$(\omega - \vec{V}_c \cdot \vec{\kappa})^2 = g\kappa'$$ (6.16)

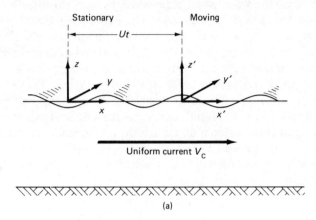

(a)

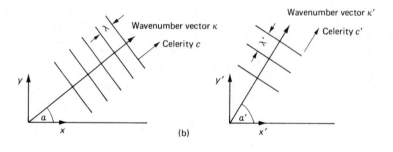

(b)

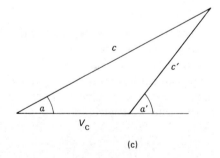

(c)

Figure 6.7 Wave modification: (a) coordinate systems; (b) wave crests; (c) wave and current speeds

So the frequency corresponding to a wave of a particular wavelength will be increased for $\vec{V}_c \cdot \vec{\kappa}$ positive, i.e. when the wave direction and current are at less than 90° to each other. These effects will modify the wave height spectrum used as an input load for our random vibration analysis. We must also remember that diffracted or reflected waves will also be modified by the current. Little or no attempt has been made to use these relations to build up a coherent theory for wave diffraction around large-diameter members in the presence of currents, but a few tentative results have been deduced; see Hogben[14]. Hogben's arguments seem to indicate that the interaction of currents with the scattered waves does not cause large increases of load at the current velocities to be expected in the North Sea, but during tow-out these effects may become important.

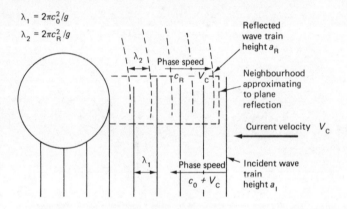

Figure 6.8   Interaction of current and diffracted waves (from Hogben[14])

To treat this problem, following Hogben, we assume that we may approximate the structure by a plane wall parallel to the wave crests of appropriate width, which reflects the incoming waves without energy loss. See Figure 6.8. For an incident wave train of height a_I and phase speed c_0 relative to and parallel to the current of velocity V_c, the height of the reflected wave train a_R is given by:

$$\frac{a_R}{a_I} = \left[\frac{2\left(1 + \dfrac{V_c}{c_0}\right)\left(1 + 2\dfrac{V_c}{c_0}\right)}{A + \sqrt{A}} \right]^{\frac{1}{2}} \tag{6.17}$$

where
$$A = 1 - 4\frac{V_c}{c_0}\left(1 + \frac{V_c}{c_0}\right)$$

Equation (6.17) was deduced using conservation of energy and wave crests with the usual linear wave theory. Energy and momentum consideration yield that the current–wave interaction modifies the wave force on the wall by a factor:

$$\frac{a_R + a_I}{2a_I} \tag{6.18}$$

Figure 6.9 shows a plot of this factor against the current speed parameter V_c/c_0 (from Hogben[14]). The critical current speed indicated by the vertical line in the diagram is that current speed which will prevent reflected waves from propagating upstream (see Taylor[5]). It should be borne in mind, however, that simple reflection is unlikely to occur under the extreme conditions near the critical current speed.

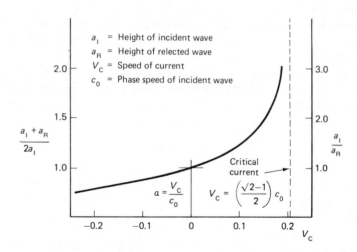

Figure 6.9 Interaction of current and diffracted waves (from Hogben[14])

The conditions of reflection assumed in this analysis are only likely to occur with members whose diameter is large when compared with the wavelength. The factor should therefore only be used for the shorter wavelengths, which in most situations do not contain much wave energy anyway. It may happen, however, during tow-out of a concrete structure, that the relative motion of the structure and the fact that the (usually large) base is on the surface may make the effects mentioned above appreciable.

6.4 SLENDER MEMBERS

Much more work has been done on the calculation of the effects of currents on slender members, since here the free-surface effects are less important, and most offshore structures before 1965 were constructed with members of this type.

The analysis of the forces on slender members is much easier because we may use Morison's equation with some success in our calculations. There has been much work done on evaluating the empirical coefficients that appear in Morison's equation, and the British Ship Research Association (BSRA) has brought out an excellent two-volume report on the subject[15]. In this report 25 recent papers are summarised, and some attempt is made to assess the results. Certain values of the two empirical coefficients c_i and c_d are suggested by the BSRA for a number of shapes and varying Reynolds number.

We are considering here the steady fluid flow past a vertical cylindrical slender member on which is superimposed an oscillatory motion due to the waves. We dealt with wave forces in Chapter 4, and the value of the Keulegan–Carpenter number, given by

$$N_{KC} = V_m T/D \qquad (6.19)$$

was considered to be the best criterion for determining the nature of the flow in purely oscillatory fluid motion. Here V_m is the maximum horizontal particle velocity, T is the period of the wave, and D is a typical dimension of the obstacle.

Keulegan and Carpenter themselves (reference 5 of Chapter 4) proposed that a wake is formed for $N_{KC} \geqslant 15$. Now we have to take into account a superimposed steady flow; the relevant measure here is the Reynolds number Re defined by:

$$Re = VD/v \qquad (6.20)$$

where V is the total fluid particle velocity, and v is the kinematic viscosity of the fluid (for water $v \simeq 0.011 \, \text{cm}^2/\text{s}$).

For the purposes of discussion the magnitude of Re is assumed to fall into three ranges typified by the nature of the flow pattern:

● the subcritical regime $Re \leqslant 5 \times 10^4$, where the flow is essentially laminar;

● the critical regime $5 \times 10^4 \leqslant Re \leqslant 5 \times 10^5$, where flow separation and reversal occur near the cylinder and an orderly 'vortex street' develops behind the cylinder;

● the postcritical regime $Re \geqslant 5 \times 10^5$, where the flow behind the cylinder becomes turbulent and disordered; there is often a reduction of drag as this regime is entered.

The effect of surface roughness

If a structure is to remain in the sea for a number of years it is likely that it will be colonised by a number of marine organisms (see Figure 6.10). These organisms may increase the weight of the structure by up to 60 kg/m² for growths of up to 200 mm in thickness. The diameter of

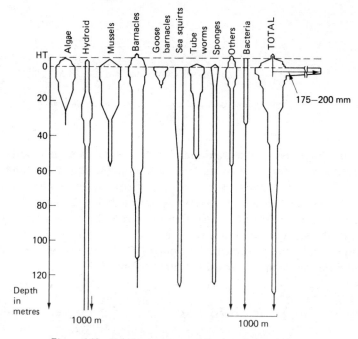

Figure 6.10   *Relative importance of various fouling organisms*

the members will be correspondingly increased and the surface roughness altered. The organisms may also have a weakening effect on the structure, as they may remove any protective coatings.

Roughness is considered to affect only the drag and hence the coefficient c_d. At subcritical Reynolds number, roughness is considered to have no effect. Increasing the roughness will reduce the critical Re at which flow separation will occur, so postcritical values of c_d must be used for lower flow velocities.

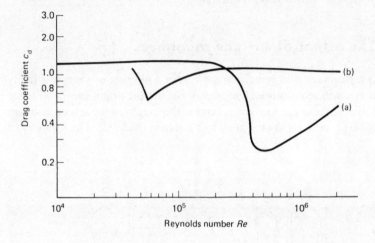

Figure 6.11   Variation of drag coefficient with Reynolds number: (a) smooth circular cylinder, (b) rough circular cylinder

For roughness of the 'sand-grain' type $c_d = 1.0$ is a good value, whereas with roughness of the 'spherical' type the equivalent sand-grain roughness is about 60 per cent of the spherical roughness and $c_d = 0.8$ should be used. Experiments on roughened cylinders have been carried out at the National Physical Laboratory[16]. Figure 6.11 shows the variation of drag coefficient with Reynolds number for a smooth and a roughened circular cylinder. Curve (a) is from Delany and Sorensen[18] and curve (b) is from Achenbach[19].

Currents and drag forces

We saw in Chapter 4 that for a slender cylinder we may write the force per unit length in the x direction at depth z as:

$$F(z,t) = \overset{\text{Inertia}}{C_I \dot{v}_x} \overset{\text{Drag}}{+ C_D v_x |v_x|} \tag{6.21}$$

where x is the direction of wave advance

v_x is the water particle velocity due to the waves in the x direction evaluated at the axis of the cylinder in its absence

C_I and C_D are empirically derived coefficients

$\dot{v}_x$ is the time derivative of v_x

This is Morison's equation. We may rewrite it in the full vectorial form as:

$$\vec{F}(z, t) = C_I \dot{\vec{V}} + C_D \vec{V} |\vec{V}| \tag{6.22}$$

Now:

$$\vec{F} = (F_x, F_y, F_z) \tag{6.23}$$

and

$$\vec{V} = (V_x, V_y, V_z) \tag{6.24}$$

(in this case $F_y = F_z = V_y = V_z = 0$), where the modulus sign may now be taken to mean the modulus of the vector $\vec{V}$ defined by:

$$|\vec{V}| = [V_x^2 + V_y^2 + V_z^2]^{\frac{1}{2}} \tag{6.25}$$

If we now have a steady uniform horizontal current velocity $\vec{V}_c$ at an angle α to the x axis and superimposed on the flow given by

$$\vec{V}_c = (V_c \cos \alpha, V_c \sin \alpha, 0) \tag{6.26}$$

we would expect to be able to simply add this velocity to $\vec{V}$ to obtain the total fluid flow velocity in our equation, giving:

$$\vec{F}(z, t) = C_I (\dot{\vec{V}} + \dot{\vec{V}}_c) + C_D (\vec{V} + \vec{V}_c) |\vec{V} + \vec{V}_c| \tag{6.27}$$

This is considered to be admissible when the angle of inclination α is not too close to 90°, representing flow transverse to the wave direction. For transverse flows the effect of the current on the waves themselves cannot be ignored and the relations presented in section 6.3 must be used.

It should be noted that for steady currents $\dot{\vec{V}}_c = 0$, and equation (6.27) becomes:

$$F(z, t) = C_I \dot{\vec{V}} + C_D (\vec{V} + \vec{V}_c) |\vec{V} + \vec{V}_c| \tag{6.28}$$

The first term representing the inertia force is hence unaffected by the presence of the current. This effect has been borne out by experiment.

Equation (6.28) may be written in component form as:

$$F_x(z, t) = C_I \dot{V}_x + C_D (V_x + V_c \cos \alpha)|\vec{V} + \vec{V}_c| \tag{6.29}$$

$$F_y(z, t) = C_D V_c \sin \alpha |\vec{V} + \vec{V}_c| \tag{6.30}$$

$$|\vec{V} + \vec{V}_c| = [(V_x + V_c \cos \alpha)^2 + V_c^2 \sin^2 \alpha]^{\frac{1}{2}} \tag{6.31}$$

It can be seen that a small current will significantly affect the direction and magnitude of the force given by Morison's equation, especially at depths greater than half a wavelength.

Caution should be used, however, when we use these equations, as it is now thought that the coefficient C_D should be taken as dependent on the current velocity, and a straightforward application of Morison's equation, as above, will give us a force that is too large (see Dalrymple[17]).

Vortex shedding

When we have steady flow of a fluid past a vertical member, vortex shedding will occur as the Reynolds number, defined by equation (6.20), increases to the value at which the flow becomes critical, in the sense that vortices will be shed. This occurs typically for $Re \geqslant 100$. Figure 6.12 shows the vortices produced behind a rectangular obstruction, obtained numerically, using finite elements[20].

Such vortex shedding occurs when the boundary layer on the obstruction separates from the surface, giving rise to reversed flow downstream of the separation point. As a result a vortex, or a pair of vortices, is formed. At low Reynolds number the vortices so formed remain behind the obstruction; this situation is shown in Figure 6.13, for $Re = 40$. As the flow velocity increases the vortices may detach themselves from the member, usually from alternate sides at a definite frequency. As a vortex detaches itself it leaves behind an equal and opposite circulation of fluid round the obstacle, similar to the starting vortex around an aircraft wing. As a result the obstacle experiences a lift force in a direction perpendicular to the flow. Usually such vortices are shed from alternate sides of the member at a frequency determined, in steady flow, by the Strouhal number, defined by:

$$S = Df/V_c \tag{6.32}$$

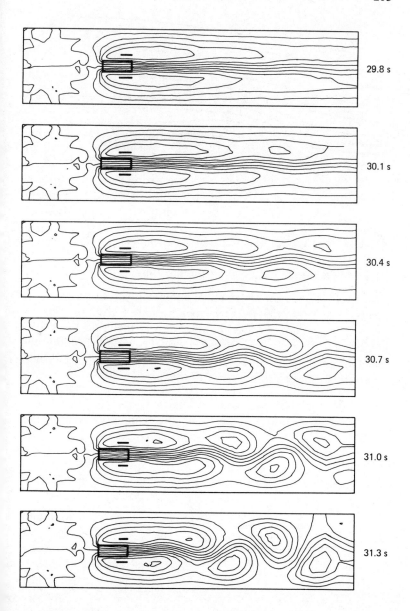

29.8 s

30.1 s

30.4 s

30.7 s

31.0 s

31.3 s

Figure 6.12   Vortex street development for Re = 100, showing stationary streamlines (which are obtained by subtracting from the streamlines those corresponding to flow without the obstacle)

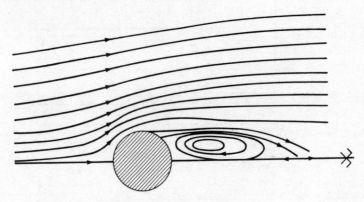

Figure 6.13 Streamlines (Re = 40)

where **D** is a typical dimension of the member
V_c is the current flow velocity
f is the frequency of vortex shedding

Figure 6.14 shows how the Strouhal number S varies with Reynolds number Re. We see that as the Reynolds number, given by equation (6.20), increases, the Strouhal number dips at the critical Reynolds number and thereafter increases. Hence the vortex shedding and resulting oscillating force will increase in frequency, and may approach the resonant frequency of the member.

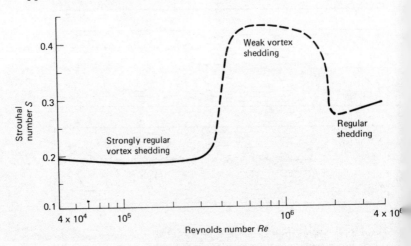

Figure 6.14 Strouhal number versus Reynolds number for smooth stationary circular cylinder

If we suspect that vortex shedding may occur, as a first approximation we may postulate that the force component due to lift F_L would be of the following form:

$$F_L(z, t) = C_L |V_c| V_c \exp(\pi i f t) \qquad (6.33)$$

where V_c is the fluid particle velocity due to the current only
f is the vortex-shedding frequency, given by equation (6.32)
C_L is an empirical coefficient dependent on the shape of the obstacle, and can be written:

$$C_L = \tfrac{1}{2}\rho D c_\ell \qquad (6.34)$$

where c_ℓ is a dimensionless number
ρ is the density of water

F_L is now perpendicular to the direction of the current and may be in a different direction from the lift force mentioned in equation (4.27). This force, due to the wave motion, is in a direction perpendicular to the direction of wave advance and is of a different nature, as we see from the equation.

The coefficient c_ℓ may be taken to be 0.7 for most calculations at low Reynolds number, although values of up to 1.5 have been recorded by Chang[21] in measurements of oscillatory flow. Figure 6.15

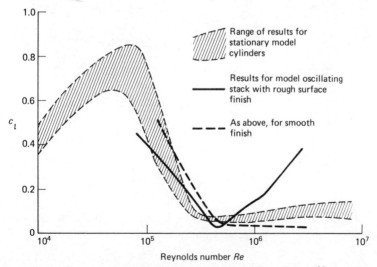

Figure 6.15   Variation of c_ℓ with Reynolds number (from Schmidt[23], courtesy American Institute of Aeronautics and Astronautics)

shows the variation of the c_f with Reynolds number (and hence flow velocity). There is a fairly sharp dip in c_f for critical values of Re, which reflects the fact that vortices are not produced in any regular fashion for these values of Re. At critical values of the Reynolds number from 4×10^5 to 8×10^5, Schmidt[23] found no predominant frequency in the lift forces; the flow is here turbulent with no clear vortices. The vortex wake is, however, re-established for very high values of Re, although this is not clear from the figure. These results are taken from Wootton et al.[22]

It should be borne in mind that the above analysis assumes a definite constant eddy-shedding frequency and a rigid stationary obstacle. If the member vibrates it is possible that the eddies may be triggered by the vibration, which will be produced at or near the resonant frequency of the member, and (as Laird[24] points out) drag and lift forces of up to four or five times those anticipated on a rigid structure may be experienced. Figure 6.16 from reference 23 shows the effect that cylinder motion has on the eddy-shedding frequency and hence the observed Strouhal number. We see that regular shedding from a freely moving cylinder continues even in the critical Reynolds number range.

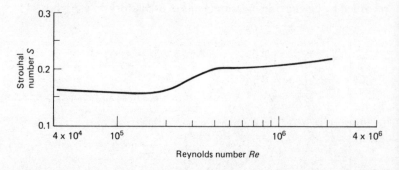

Figure 6.16   Strouhal number versus Reynolds number for cylinder free to oscillate (from Schmidt[23], courtesy American Institute of Aeronautics and Astronautics)

Below we present an illustrative calculation of the forces on a vertical circular cylinder in a uniform current of velocity V_c.

Example 6.1

Consider a circular wooden pile that has been driven into a river bed. The length of the pile and depth of the river l is 20 metres, the diameter D is 0.40 m. Compute the drag forces in the pile, assuming the velocity of the water V_c is constant, and determine the value of V_c for which vortex shedding occurs. The following values apply:

Modulus of elasticity of wood $E = 10^{10}\,\text{N/m}^2$
Density of wood $\rho = 0.8 \times 10^3\,\text{kg/m}^3$
Density of water $\rho_w = 10^3\,\text{kg/m}^3$
Moment of inertia $I = \pi R^4/4$ (radius $R = D/2$)

The drag forces can be computed using the following formulae ($c_d = 1$ from Figure 6.11):

$$F_D = \tfrac{1}{2}\rho_w D c_d V_c |V_c|$$
$$= 2 \times 10^2 V_c |V_c| \qquad \text{(a)}$$

The fundamental frequency of the pile can be obtained using Rayleigh's quotient:

$$\omega_n^2 = \frac{EI \displaystyle\int_0^l \left(\frac{\partial^2 f^2}{\partial x^2}\right) dx}{\rho_w A \displaystyle\int_0^l f^2\,dx + C_M \displaystyle\int_0^l f^2\,dx} \qquad \text{(b)}$$

where $C_M = c_m \cdot D^2 \rho_w \pi/4$ and $c_m = 1$. This term involving C_M is due to the hydrodynamic mass.

In order to solve (b) let us assume that the shape of the pile can be given as $f = \bar{x}^2$, where $\bar{x} = x/l$. Hence:

$$\omega_n^2 \simeq \frac{\dfrac{4EI}{l^3}}{\dfrac{\rho A l}{5} + c_m \dfrac{D^2 \rho_w \pi}{4}\dfrac{l}{5}} \simeq 10\,\text{s}^{-1} \qquad \text{(c)}$$

Thus circular frequency $\omega_n \simeq 3.16\,\text{s}^{-1}$ and frequency $f_n \simeq 0.5\,\text{s}^{-1}$.
The Strouhal number for the cylinder at high Re numbers is:

$$S = f_s D/V_c \simeq 0.2$$
$$f_s = 0.2 V_c/D \qquad \text{(d)}$$

Note that in order for f_s to equal f_n we need a velocity of:

$$V_c = f_n\,D/0.2 = 1\,\mathrm{m/s} \qquad (e)$$

At this velocity there is a danger that the pile will break because of vortex shedding. The calculation is approximate, owing to the modal shape we have assumed. It can be improved by taking a better shape or by dividing the pile into a number of beam elements.

Interaction between members

If one member is situated in front of another with respect to the current direction, its wake will impinge on the other. This effect will depend on the Reynolds number of the flow around the first member and the separation between the members. For low-Reynolds-number flow around the first of two cylinders there will be little or no interaction, if the downstream member is outside the two attached vortices on the first cylinder. The direction of flow around the second cylinder may, however, be changed, thus altering the direction of drag and lift forces on this cylinder—Figure 6.17(a).

As the flow speed increases, vortices may now be shed from the first cylinder and disturb the flow around the second—Figure 6.17(b). This effect will continue to be important even when no periodic vortices are shed, as the wake will contain a shear-velocity profile that will induce lift forces on the second cylinder—Figure 6.17(c).

Figure 6.17(a) shows that the force on the second cylinder may be calculated in the usual way using the *local flow* velocities in the Morison equation for this member. In the situation of Figure 6.17(b), however, the disturbed incident flow on the second cylinder will have two effects:

● The drag force on the second cylinder will be reduced.

● The circulation of an incident vortex will subsequently induce a circulation around the cylinder, and induce a lift force in the appropriate direction: in the diagram a clockwise vortex will induce an upward force and an anticlockwise one a downward force. Some information on the magnitude of the vortices and the proportion of this rotational motion, which is converted to circulation around the second cylinder, is needed to calculate the magnitude of the lift forces involved.

The situation shown in Figure 6.17(c), in which a strongly turbulent

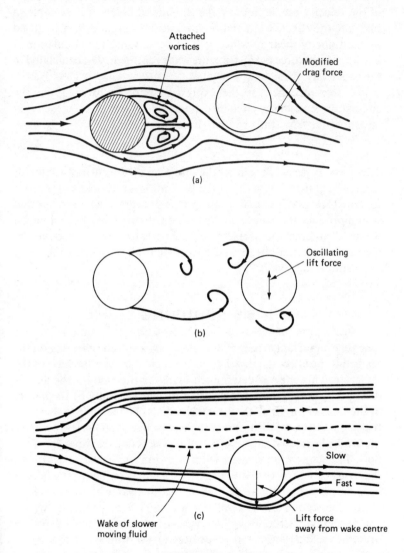

Figure 6.17 Interaction effects at different Reynolds numbers: (a) flow direction modification; (b) vortex interaction; (c) very high Re

wake has been formed, is still more difficult to analyse. Some measure of the velocity profile in the wake is required before it is possible to calculate the effective circulation around the second cylinder caused by the velocity shear in which it is situated. Once the circulation is known, the lift force (away from the wake centre) may be calculated in the usual way.

For reference here is the expression for the force, given the circulation κ round a circular cylinder in a current of velocity V_c measured far from the cylinder (Batchelor[25]):

$$F = \rho V_c \kappa \tag{6.35}$$

This force is perpendicular to the direction of the current (upwards for a current from left to right); the circulation κ is considered positive for anticlockwise flow, and is equal to the strength of a line vortex that would produce the same velocity field. Little work has been done on the wake interaction problem, as most treatments have looked at inertial interaction effects and have specifically excluded the wake (Yamamoto and Nath[26]).

6.5 SCOURING

One very important aspect of fluid flow past a structure resting on the sea bed is the effect the flow has on the region around the base of the structure. This effect is determined firstly by the amount the fluid is diverted by the presence of the structure, and secondly by the nature of the sea bed on which the structure stands. The larger the structure the larger will be the disturbance to the flow field, and the less cohesive the sea bed material the greater will be the scour resulting from the flow. The scour will originate from two sources: the wave action on the base of the structure, and the effect of currents. The first will only become important for structures in shallow water or for very long waves; the second will generally be a larger effect, particularly with concrete gravity structures whose bases can be as much as 100 metres in diameter. Steel lattice-type structures, on the other hand, are kept in place by piling, and the diameters of the base members are typically much smaller, so that scour effects are of a much lesser importance.

Wave-induced scour may in fact be important because of its cyclic nature, as the oscillating wave force on the top of the structure may cause the sea bed under the foundation to weaken (this topic is dealt with more fully in Chapter 9).

The mechanism of scour

In order to understand the mechanism of scour it is necessary to study the flow field around the base of a typical marine structure. We shall take, as a prototype example, a vertical circular cylinder resting on the sea bed. This situation has most of the qualitative features of flow that will occur around any blunt body.

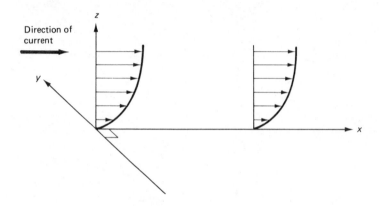

Figure 6.18   Velocity profile near sea bed for uniform current

Because of the small but finite viscosity of water, the velocity profile of the flow near the sea bed due to a uniform current will be approximately as in Figure 6.18. If we now draw in the lines of constant vorticity we obtain Figure 6.19. These vortex lines represent the shear flow, their proximity representing the amount of circulation in a particular region; it can be shown that they will move largely with the fluid.

Figure 6.20 shows the effect that a blunt obstacle will have on the vortex lines, and the corresponding circulation of fluid that will occur at the base of the obstacle where the velocity gradient is greatest. It can be seen that a strong region of circulation is set up on the upstream part of the body and that two trailing vortices continue downstream of the obstacle. At higher Reynolds number a third, often turbulent,

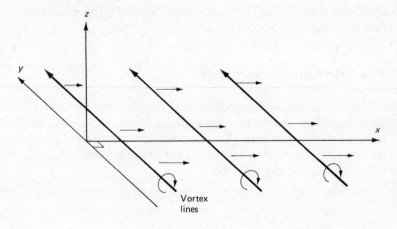

Figure 6.19   Vortex lines for uniform current with bottom friction

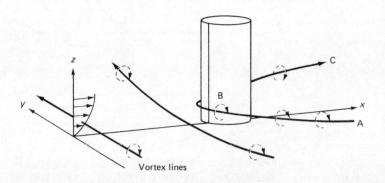

Figure 6.20   Flow past circular cylinder (scour will occur mainly below vortex line ABC)

vortex system will develop directly behind the body, causing additional scour in this region.

We may now expect that, in the regions of strong circulation near the sea bed, particles will be lifted into suspension and subsequently swept away with the flow. This process continues until an equilibrium condition is reached in which the sea bed assumes a form less conducive to vortex formation, and natural infilling from the surrounding area compensates the scouring process. A number of papers have been written on the quantitative aspects of this topic,

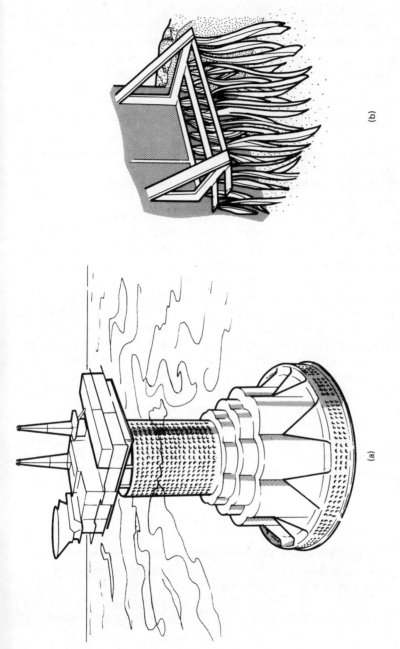

Figure 6.21 Scour prevention: (a) perforated screen, (b) fibre curtain

notably Machemehl and Abad[27] and Palmer[28]. Since the main approach adopted by designers is to prevent the occurrence of scouring, the quantitative aspects of the phenomenon will be omitted here and we shall concentrate on the methods of prevention commonly employed.

Scour prevention

There are two methods of reducing scouring round an offshore structure.
- Design the base of the structure so that the resulting flow field does not produce appreciable circulation.
- Change the nature of the sea bed around the structure.

There have been a number of designs that utilise the first method. The most spectacular was used for the central platform for the Ninian field; see Figure 6.21(a). The base of this structure is surrounded by a cylindrical circular perforated screen, which by allowing the water to circulate behind the screen at a lower velocity induces any suspended particles to be deposited behind the screen; in this way scouring is prevented. Another solution based on the same idea of slowing down the flow is illustrated in Figure 6.21(b). In this system a curtain of Terylene fibre hung around the base induces the build-up of a stable sandbank; this method has been used successfully in the Frigg oil field.

The second, more obvious, method has been more widely used. It involves dumping rubble or concrete blocks around the structure when it is in place. This may be done from barges, or the rubble may be released from the structure as it is positioned by means of hinged flaps attached to its base. Various other preparations to the sea bed may be made before the structure is positioned.

6.6 WIND CHARACTERISTICS

In order to calculate the wind forces on an offshore structure it is necessary to have some information on the characteristics of the wind climate at that particular location.

Wind direction and mean velocity

The wind direction and velocity are usually represented by a 'wind rose' such as the one shown in Figure 3.12. The directions of the radial lines represent the directions of the expected winds. The thickness of the line represents the strength of the wind, and the length at a certain thickness represents the percentage of the total time we would expect a wind of that strength in that particular direction. Since most structures are asymmetrical a number of trial calculations should be made for the varying directions; the structure may then be oriented to minimise the expected wind loads. For offshore structures this may not be possible, however, as the predominant wave and current directions may be the deciding factor for a decision of this kind. This is often the case: typically the wind loads are less than 10 per cent of the total loads on an offshore structure. A more sophisticated statistical analysis (used for fatigue problems, for example) may be needed. In this case the maximum wind speeds likely to occur in a specified return period may be required. The most common source of such data is the Meteorological Office; see reference 29.

The designer must also take into account the vertical velocity profile of the wind as it travels over the water surface. It is clear that large waves will deflect the air flow (see Figure 3.11), and this may become important, particularly in the region between the waves and the deck. For details of these effects see Davidson and Frank[30] and Shemdin[31].

We now neglect the effect of the waves on the wind profile and use the commonly accepted law for the mean wind velocity profile over open terrain. If we define the gradient height Z_G as the height above (mean) sea level at which the mean wind speed attains an almost constant velocity, usually called the gradient velocity $\overline{V}_G$, we may calculate the mean wind velocity at any height Z by:

$$\overline{V}(Z) = \overline{V}_G \left[\frac{Z}{Z_G} \right]^{\alpha} \tag{6.36}$$

For open sea we take $\alpha = 0.16$ and $Z_G = 300$ metres. The above relation was suggested by Davenport[32].

226

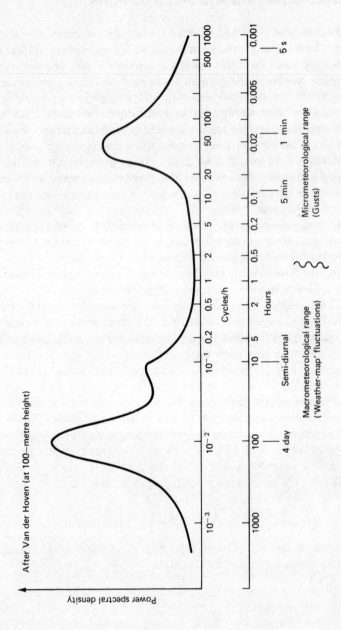

Figure 6.22 Power spectral density of fluctuating wind velocities

Fluctuating wind velocity component

In addition to the mean wind velocity there will also be a fluctuating component that must be taken into account when calculating the response of flexible members of a structure. This fluctuating component must be dealt with statistically, the most useful measure being the power spectral density of the wind velocities. Figure 6.22 shows such a spectrum derived by Van der Hoven from a variety of wind records. It is clear from the figure that there is a large amount of energy in the fluctuations with time period 5 seconds to 5 minutes; air motions of this type are classified as gusts and may be important for calculating the forces on the tower used for stacking the drill pipe sections. We also see that there is little wind energy between 10 minutes and 2 hours, which helps us to obtain time-averaged wind velocities. An averaging period of one hour has conventionally been used in most countries, although a shorter period of about 15 minutes might be more appropriate, to take into account the more sudden gusts that occur.

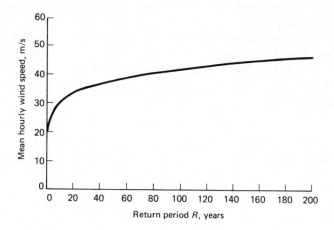

Figure 6.23   Return-period curve

When we have obtained values of mean hourly wind velocities we need in some cases to perform extreme data analysis to arrive at design wind speeds associated with given probabilities. The return period R is related to $P(< \overline{V})$, the probability that a certain wind velocity is not exceeded in any year, by:

$$R = \frac{1}{1 - P(< \overline{V})} \qquad (6.37)$$

The return period is the average period between occasions on which velocity $\overline{V}$ is reached or exceeded. Figure 6.23 shows a typical return period curve.

The probability that $\overline{V}$ is not exceeded in N consecutive years is $P(< \overline{V})^N$, and the probability that it is exceeded is:

$$P_N(< \overline{V}) = 1 - [P(< \overline{V})]^N \qquad (6.38)$$

and so $$P_N(< \overline{V}) \simeq N/R \qquad (6.39)$$

For a structure of expected (useful) lifetime of N years the 'risk' N/R may then be calculated from the design wind speed.

Turbulence

The fluctuating component of the wind will have three components, two perpendicular to the mean wind direction and the third in line with the wind. It has been shown theoretically that the turbulence decays with height, but measurements show it to be reasonably constant up to most heights of interest over open country or a moderate sea. Davenport and Harris have deduced the important result that

$$\sigma_v = 0.11 \, \overline{V}_G \qquad (6.40)$$

where v is the fluctuating component of the velocity and σ_v is the standard deviation of this component. Near the surface the in-line component is much larger than the other two. We shall confine ourselves to consideration of this component.

Von Karman and Davenport[34] have proposed the following form for the power spectral density of the in-line velocity component:

$$S_{vv}(f) = \frac{4kf\tilde{V}_{10}^2}{(2 + \tilde{f}^2)^{5/6}} \qquad (6.41)$$

where f is the frequency
    $\tilde{f}$ is a normalised frequency, given by $\tilde{f} = fL/\overline{V}_{10}$
    k is the surface drag coefficient (0.005 for open sea)
    $\overline{V}_{10}$ is the mean wind velocity at 10 m height

L   is a representative-length scale for the turbulence, taken to be 1200 m by Davenport

This curve is shown in Figure 6.24.

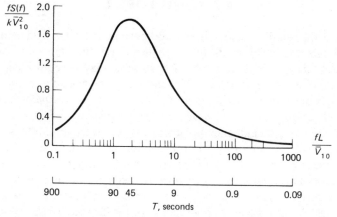

Note: T corresponds to a wind of $\bar{V}_{10}$ = 20 m/s and L = 1800 m

Figure 6.24   Universal spectrum of horizontal turbulence (gusts)

We now have enough information to calculate wind velocities in a statistical sense at any point, but to calculate the forces on extended areas of a structure we need some measure of the coherence of these velocities, so that we may combine the forces on each part of the structure to get some kind of response function. Such a measure is the coherence function linking the two points x_1 and x_2, defined by:

$$\gamma_{x_1 x_2}(f) = \frac{\left|S_{x_1 x_2}(f)\right|}{\left[S_{x_1}(f) S_{x_2}(f)\right]^{\frac{1}{2}}} \qquad (6.42)$$

A good approximation to this function is given by:

$$\gamma_{x_1 x_2}(f) = \exp\left[-\frac{Cfr}{\bar{V}(Z)}\right] \qquad (6.43)$$

where r is the distance between the two points. For vertical separation we may put $C = C_V = 9$; for horizontal separation $C = C_H = 16$ is a reasonable value.

C depends on the particular surrounding obstacles; the above values are quoted for a plane solid wall facing the wind. See equation (6.58) for a rigorous definition of this function.

6.7 WIND LOADS

Once the wind velocities have been determined we may begin to estimate the wind loads to be expected on the structure. As a preliminary estimate, records of past work may be consulted, for example Aquisne and Boyce[35]. Below is a review of the present techniques for calculating these forces on the various *parts* of the superstructure; however, because of the complicated geometry of the various components, and the obvious space restriction, interaction effects will be important. For example, the positioning of living quarters may make the use of the helicopter landing pad impossible in certain wind conditions, so care must be taken in choosing the relative positions of the components on the deck. For this reason a wind tunnel test on a model is almost essential for detailed examination of this problem. The results from such a test will not be too unreliable, as the Reynolds number of the airflow does not significantly affect the flow over sharp-edged bodies (although it *is* important for flow around large cylindrical bodies such as the legs of a concrete gravity-type structure during tow-out). It should be borne in mind that, if the structure is to operate in arctic conditions, ice accretion may occur on the superstructure and substantially affect its dynamic and drag characteristics.

We shall now review the techniques for calculating the forces on the components of the superstructure, ignoring interaction effects.

Geometrical classification

Figure 6.25 shows in diagrammatic form the layout of the superstructure of a typical offshore platform, with the different components classified according to their geometrical type, which will determine their mode of response to wind forces.

● Type 1—drilling tower and hoist equipment, consisting mainly of cylindrical members of small diameter. The flow is largely determined by the Reynolds number and the forces calculated using a Morison

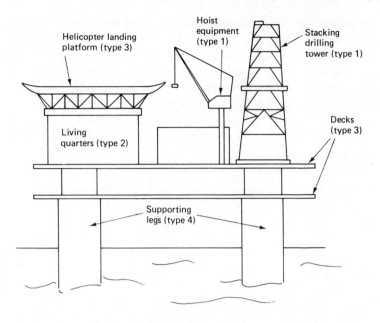

Figure 6.25 Superstructure of typical offshore platform, showing geometry types

type equation. Only the drag term of the above is important, because of the relatively low density and high compressibility of the air. Vortex shedding may occur causing vibration transverse to the flow direction.

● Type 2—living quarters, offices etc., consisting of rectangular surfaces. The flow is distorted in a three-dimensional way and separates at the sharp corner, giving rise to a wake region and eddies at these locations. The cross-correlation between the shedding forces is small and may often be ignored. The forces are calculated using Davenport's gust factors for a plane vertical wall facing the wind.

● Type 3—decks and helicopter landing platform, consisting largely of flat horizontal surfaces edge on to the wind. The flow is parallel to the surface, and a separation vortex may form at the leading edge, giving rise to flutter. The main force on the upper surface is predominantly upward, which may give rise to vertical motion of this type of structure.

● Type 4—platform supporting columns. During tow-out of a gravity platform, these components present the same problems of wind loading as a group of power-station cooling towers. Since they are essentially shell-like structures, the analysis must use full spectral analysis with certain cross-correlation assumptions depending on the tapering and vertical structure of the wind velocity profile. The exposed length of these columns during tow-out may be as much as 130 metres with a base diameter of 20 metres. In these circumstances a Morison-type approach is not appropriate.

Type 1 structures

For structures consisting of a number of cylindrical members of small diameter that do not appreciably disturb the flow, we may write the wind pressure at a point as:

$$P(t) = \tfrac{1}{2}\rho_a c_d V^2(t) \tag{6.44}$$

where $P(t)$ is the time-dependent pressure

ρ_a is the density of air

c_d is a drag coefficient that depends on the geometrical shape; as a first approximation we may use the values given earlier (see Figure 6.11)

$V(t)$ is the time-dependent wind velocity

Splitting the pressure and velocity into mean and fluctuating components

$$V(t) = \bar{V} + v(t) \tag{6.45}$$

$$P(t) = \bar{P} + p(t) \tag{6.46}$$

and neglecting terms of order $v(t)$, we may obtain:

$$\bar{P} + p(t) = \tfrac{1}{2}\rho_a c_d \bar{V}^2 + \rho_a c_d \bar{V}v(t) \tag{6.47}$$

or $$\bar{P} = \tfrac{1}{2}\rho_a c_d \bar{V}^2 \tag{6.48}$$

and $$p(t) = \rho_a c_d \bar{V}v(t) \tag{6.49}$$

The pressure and velocity spectra are then related by:

$$S_{pp}(t) = (c_d \rho_a \bar{V})^2 S_{vv}(f) \tag{6.50}$$

$$= (2\bar{P}/\bar{V})^2 S_{vv}(f) \tag{6.51}$$

Water flow round a circular cylinder may produce lift forces, as described in section 6.4. In the same way the airflow may separate and air vortices may be shed, giving rise to a lift force at the shedding frequency f_s given by the Strouhal number S. We have:

$$S = Df_s / \overline{V} \qquad (6.52)$$

and the force may be written approximately as:

$$F(z, t) = \tfrac{1}{2}\rho_a Dc_\ell \,|\,\overline{V}\,|\,\overline{V}\exp(\pi i f_s t) \qquad (6.53)$$

S has a value near 0.16 for a circular cylinder. For design purposes, values of the lift coefficient of up to 1.0 are possible, the high values corresponding to sub-critical conditions.

Equation (6.44) is only strictly true when the dimensions of the obstruction are much smaller than the turbulence wavelength defined by $\lambda = \overline{V}/f$. For type 2 structures this may not be true.

Type 2 structures

For buildings whose dimensions are comparable with the turbulence wavelength we must include an extra term in equation (6.51). This term is usually referred to as the 'aerodynamic admittance function' χ^2. We therefore have:

$$S_{pp}(f) = (2\overline{P}/\overline{V})^2 \chi^2 S_{vv}(f) \qquad (6.54)$$

Vickery[36] has suggested the following expression for χ:

$$\chi = \frac{1}{1 + (2f\sqrt{A}/\overline{V})^{4/3}} \qquad (6.55)$$

where A is the area of the building presented to the airflow. This function is plotted in Figure 6.26 for various values of $f\sqrt{A}/\overline{V}$.

If we integrate equation (6.54) over the exposed area of the building to obtain the force spectrum, we shall have assumed total correlation in the turbulence at different points. We must therefore utilise the coherence function given in equation (6.43) to take the effect of phase variation into account. A full analysis of this problem can be found in Davenport[37].

The preliminary expression for the generalised pressure spectrum expressed as a function of the two points (y_1, z_1) and (y_2, z_2) in the

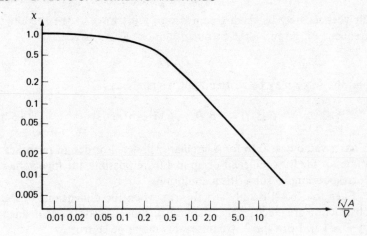

Figure 6.26   Graph of aerodynamic admittance function

plane perpendicular to the mean wind direction is given by:

$$S_{pp}^r(y_1, y_2, z_1, z_2, f) = 4\exp\left(-\frac{C_V rf}{\overline{V}_G}\right)\exp\left(-\frac{C_H rf}{\overline{V}_G}\right) \times$$

$$\times \frac{\overline{P}_G{}^2}{\overline{V}_G{}^2}\chi^2\frac{\overline{V}_1\,\overline{V}_2}{\overline{V}_G{}^2}S_{vv}(f) \qquad (6.56)$$

where the subscripts 1 and 2 denote that the variable is to be evaluated at the corresponding points. The superscript r indicates that this spectrum corresponds to the rth mode of vibration.

This spatial spectrum is linked to the spectrum of the generalised force for a mode r of the motion of the exposed face of the building by:

$$S_{pp}^r(f) = \iiiint\limits_{\text{over area}} S_{pp}^r(y_1, y_2, z_1, z_2, f)\phi_r(y_1, z_1) \times$$

$$\times \phi_r(y_2, z_2)\,\mathrm{d}y_1\,\mathrm{d}y_2\,\mathrm{d}z_1\,\mathrm{d}z_2 \qquad (6.57)$$

where $\phi_r(y, z)$ is the mode shape corresponding to the rth mode. Equation (6.57) leads to a definition of the coherence function γ mentioned in equation (6.42), i.e.

$$S_{pp}^{r}(f) = S_{pp}(y_0, z_0, f)A^2 \iint \gamma(y_1, z_1, y_2, z_2, f) \times$$

$$\times \phi_r(y_1, z_1)\phi_r(y_2, z_2)\frac{dy_1\,dy_2\,dz_1\,dz_2}{A^2} \qquad (6.58)$$

where (y_0, z_0) is some reference point at which the spectrum $S_{pp}(y_0, z_0)$ is evaluated.

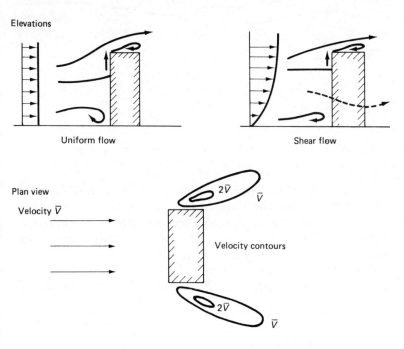

Figure 6.27   Flow around buildings (from Maccabbee[38])

At the sharp corner of the building, flow separation may occur and lateral forces may build up in the usual way. These may be calculated using a lift coefficient as before (see equation (6.53)) if one is available, but because of the three-dimensional structure of the flow the forces calculated in this way may be highly inaccurate. A qualitative picture of the type of flow to be expected is shown in Figure 6.27. Note the separation point on the upstream side of the building and the downflow around the sides in shear flow.

Type 3 structures

For flat horizontal surfaces edge on to the wind the main problem is vertical deflection of the structure, which may cause flow separation as in Figure 6.28. The flow separation may redistribute the pressure, and energy may then be transferred to the structure by this process. When

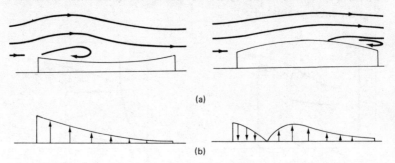

Figure 6.28   Type 3 structure: (a) flow patterns over suspension roof; (b) mean pressures on upper surfaces

a regular cycle of deflections is set up, extracting energy from the wind, the coupled motion is called flutter. This does not often occur in the situations we are concerned with, because the stiffness is often too high and the plates out of which decks are usually constructed are made in a rectangular grid fashion; this allows air to pass through, giving pressure equalisation on the two sides. The interested reader is referred to books and papers on the dynamics of suspension roofs and bridges.

Type 4 structures

The large cylindrical supporting legs of a gravity structure must be analysed as shells of revolution responding to random loads, because of their enormous size and varying cross-section.

A full generalised-force spectral analysis must be used, as was indicated in equation (6.57) for rectangular buildings. We take cylindrical polar coordinates and choose mode shapes corresponding to an integral number of circumferential waves, n, and meridional shapes in the z direction (see Figure 6.29). Deflections in the z direction result in distortions of the cross-section. If this effect is neglected a simple beam analysis may be used.

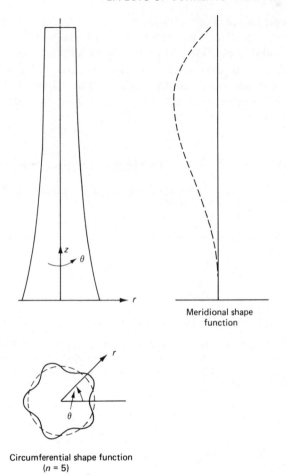

Meridional shape
function

Circumferential shape function
$(n = 5)$

Figure 6.29   Deflection of supporting column of offshore gravity platform

A diffraction theory analysis is unfortunately not appropriate, because the lower density and greater compressibility of the air gives rise to a wake behind each column, within which the pressures are of a qualitatively different nature. To deal with this, the surface of each column is divided into three independent sets of strips:

- vertical
- front circumferential
- rear circumferential

The coherence functions between the members of each of these sets have been calculated by Abbu-Sitta and Hashish[39, 40] and the corresponding deflections calculated for hyperbolic cooling towers.

Because of the (possible) importance of the interaction effects between the columns, neglected in the currently available literature, we do not propose to go into the complex analysis involved in the above problem.

References

1. Ekman, V. W., 'Meeresströmungen', *Hanb. Phys. Techn. Mechanik*, **5**, 196, Hrg. von Auerbach u. Hort, Leipzig (1927)
2. Defant, A., *Physical oceanography* (2 vols), Pergamon Press (1961)
3. Muir Wood, A. M., *Coastal hydraulics*, Macmillan (1969)
4. Longuet–Higgins, M. S., and Stewart, R. W., 'The changes in amplitude of short gravity waves on steady non-uniform currents', *J. Fluid Mech.*, **10**, 529–549 (1961)
5. Taylor, Sir Geoffrey, 'The action of a surface current used as a breakwater', *Proc. R. Soc. (Series A)*, **231**, 466–478 (1955)
6. Stoker, J. J., *Water waves*, Interscience Publishers (1957)
7. Wehausen, J. V., and Laitone, E. V., 'Surface waves', *Encyclopaedia of Physics*, Vol. 9, Springer, Berlin (1960)
8. Lunde, J. K., 'On the linearised theory of wave resistance for displacement ships in steady and accelerated motion', *Trans. SNAME*, **59**, 24–76 (1951)
9. Hogben, N. G., and Standing, R. G., 'Experience in computing wave loads on large bodies', OTC 2189, Proc. Offshore Technology Conf., Houston (1975)
10. Kotik, J., and Morgan, R., 'The uniqueness problem for wave resistance calculated from singularity distributions which are exact at zero Froude number', *J. Ship Res.*, **13**, 61–68 (1961)
11. Hogben, N., *Wave resistance of surface piercing vertical cylinders in uniform currents*, NPL Ship Div. Report 183 (1974)
12. Longuet–Higgins, M. S., and Stewart, R. W., 'Changes in the form of short gravity waves on long waves and tidal currents', *J. Fluid Mech.*, **8**, 565–83 (1960)
13. Johnson, J. W., 'The refraction of surface waves by currents', *Trans. Am. Geophys. Union*, **28**, No. 6, 867–874 (1947)
14. Hogben, N., 'Wave loads on structures', Proc. Behaviour of Offshore Structures Conf., Trondheim (1976)
15. British Ship Research Association, *A critical evaluation of the data on wave force coefficients*, Report No. W.278 (1976)
16. Miller, B. L., Maybrey, J. F., and Salter, I. J., *The drag of roughened cylinders at high Reynolds numbers*, NPL Mar. Sci. Report 132 (1975)
17. Dalrymple, R. A., 'Waves and wave forces in the presence of currents', Proc. Conf. Civil Engineering in the Oceans, Univ. Delaware (1975)
18. Delany, N. K., and Sorensen, N. E., *Low speed drag of cylinders of various shapes*, US National Advisory Committee for Aeronautics, Tech. Note 3038 (1953)
19. Achenbach, E., 'Influence of surface roughness on the cross flow around a circular cylinder', *J. Fluid Mech.* **46**, 321–335 (1971)
20. Smith, S., and Brebbia, C. A., 'Finite element solution of Navier–Stokes equations for transient two-dimensional incompressible flow', *Physics of Fluids*, **17** (1975)

21. Chang, K. S., *Transverse forces on cylinders due to vortex shedding in waves*, MSc thesis, MIT Dept Civil Engineering (1964)
22. Wootton, L. R., Warner M. H., Sainsbury, R. N., Cooper, D. H., *Oscillation of piles in marine structures*, CIRIA Tech. Note 40 (1972)
23. Schmidt, L. V., 'Measurements of fluctuating air loads on a circular cylinder', *J. Aircraft*, **2**, No. 1 (Jan. 1965)
24. Laird, A. D. K., 'Water forces on flexible oscillating cylinders', *Proc. ASCE (Waterways and Harbors Div.)*, **88**, WW3 (Aug. 1962)
25. Batchelor, G. K., *An introduction to fluid dynamics*, Cambridge U. P. (1967)
26. Yamamoto, T., and Nath, J. H., 'Hydrodynamic forces on groups of cylinders' OTC 2499, Proc. Offshore Technology Conference, Houston (1976)
27. Machemehl, J. L., and Abad, G., 'Scour around marine foundations', OTC 2313, Proc. Offshore Technology Conf., Houston (1975)
28. Palmer, H. D., 'Wave-induced scour on the sea-floor', Proc. Civil Engineering in the Oceans II, Miami Beach, 703–716 (1969)
29. Department of Energy, *Guidance on the design and construction of offshore installations*, HMSO, London (1974)
30. Davidson, K. L., and Frank, A. J., 'Wave-related fluctuations in the airflow above natural waves', *J. Phys. Oceanogr.* (Jan. 1973)
31. Shemdin, O. H., 'Wave influence on wind velocity profile', *Proc. ASCE (Waterways and Harbors Div.)*, **96**, WW4 (Nov. 1970)
32. Davenport, A. G., 'Rationale for determining design wind velocities', *Proc. ASCE (Struct. Div.)*, **86**, ST5 (May 1960)
33. Anthony, K. C., 'The background to the statistical approach', in *The modern design of wind sensitive structures*, CIRIA (1971)
34. Davenport, A. G., 'The application of statistical concepts to the wind loading of structures', *Proc. Instn Civ. Engrs*, **19** (Aug. 1961)
35. Aquisne, J. E., and Boyce, T. R., 'Estimation of wind forces on offshore drilling platforms', R. Inst. Naval Architects Spring Meeting (1973)
36. Vickery, B. J., 'On the reliability of gust loading factors', *Civ. Engng Trans. Inst. Engrs, Australia*, **CE 13**, No. 1 (Apr. 1971)
37. Davenport, A. G., 'Gust loading factors', *Proc. ASCE (Struct. Div.)*, **93**, ST3 (June 1967)
38. Maccabbee, F. G., 'The present state of knowledge of flow round buildings', Symp. on wind effects on buildings and structures, Loughborough (1968)
39. Abbu–Sitta, S. H., and Hashish, M. G., 'Dynamic wind stresses in hyperbolic cooling towers', *Proc. ASCE (Struct. Div.)*, **99**, ST9 (Sep. 1973)
40 Hashish, M. G., and Abbu–Sitta, S. H., 'The response of hyperbolic cooling towers to turbulent wind', *Proc. ASCE (Struct. Div.)*, **100**, ST5 (May 1974)

7 Response of a One-degree-of-freedom System

7.1 INTRODUCTION

In this chapter we examine the dynamic behaviour of a one-degree-of-freedom system under forced deterministic and random loading.

Consider first a simple mass–spring system as shown in Figure 7.1. The behaviour of the system can be analysed in terms of the displacement u. More complex systems will be represented by many coordinates, but by applying a transformation it is possible to work in terms of generalised coordinates, the behaviour of each being governed by an uncoupled equation. In this way each coordinate can be studied effectively as a one-degree-of-freedom system, as shown in this chapter. We shall see how to carry out this transformation in Chapter 8.

The system of Figure 7.1 will be in equilibrium when the system is at

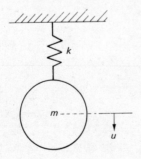

Figure 7.1   Spring–mass system

240

rest and the force in the spring k is equal to the weight of the mass (weight $= mg$, where g is the acceleration due to gravity). Hence:

$$mg = ku_s$$

$\therefore$
$$mg - ku_s = 0 \tag{7.1}$$

where u_s denotes the static displacement. If the mass is displaced a further distance u it will experience a restoring force F_a, such that:

$$F_a = mg - k(u_s + u) \tag{7.2}$$

Taking into account the static equilibrium, we have:

$$F_a = -ku \tag{7.3}$$

D'Alembert's law states that at each moment in time the restoring force will be:

$$F_a = m\ddot{u} \tag{7.4}$$

where $\ddot{u}$ is the acceleration. Here we have, from (7.3) and (7.4), that:

$$m\ddot{u} + ku = 0 \tag{7.5}$$

The solution of this equation represents the *free vibrations* of the spring–mass system, which has a harmonic solution of the type:

$$u = A \cos \omega_r t + B \sin \omega_r t \tag{7.6}$$

where A and B are constants to be determined from the initial conditions. Substituting (7.6) into (7.5) we find:

$$(-\omega_r^2 m + k)(A \cos \omega_r t + B \sin \omega_r t) = 0 \quad \forall t \tag{7.7}$$

Thus

$$-\omega_r^2 m + k = 0 \tag{7.8}$$

or
$$\omega_r = \sqrt{(k/m)} \tag{7.9}$$

ω_r is the *natural frequency of the system*.

The displacement is governed by equation (7.6), and we can now determine A and B from the initial conditions of the system (i.e. the conditions at time $t = 0$). If we define u_0 as the displacement of the system at $t = 0$ and $\dot{u}_0$ as the velocity of the system at $t = 0$, we find that:

$$A = u_0, \quad B = \frac{\dot{u}_0}{\omega_r} \tag{7.10}$$

Hence the displacement can now be written:

$$u = u_0 \cos \omega_r t + \frac{\dot{u}_0}{\omega_r} \sin \omega_r t \tag{7.11}$$

This function can be represented by a single function:

$$u = C \sin (\omega_r t + \alpha) \tag{7.12}$$

where

$$C = \left(u_0^2 + \frac{\dot{u}_0^2}{\omega_r^2} \right)^{\frac{1}{2}}, \quad \tan \alpha = \frac{u_0 \omega_r}{\dot{u}_0}$$

Function (7.12) can be plotted as shown in Figure 7.2(a). C is the amplitude of the displacements and α is the phase angle. The time between two peaks is called the period T:

$$\omega_r T = 2\pi \tag{7.13}$$

$$\therefore \qquad T = \frac{2\pi}{\omega_r}$$

The frequency f is sometimes expressed in hertz or cycles per second:

$$f = \frac{1}{T} = \frac{\omega_r}{2\pi} \tag{7.14}$$

We can also plot the velocity or derivative of (7.12), i.e.

$$\dot{u} = \omega_r C \cos (\omega_r t + \beta) \tag{7.15}$$

The initial velocity is $\dot{u}_0$; see Figure 7.2(b).

Consider now that the spring–mass system is subjected to a harmonic varying force (Figure 7.3). A simple force, for instance, is:

$$F = P \sin \omega t \tag{7.16}$$

The governing equation for the system is then:

$$m\ddot{u} + ku = P \sin \omega t \tag{7.17}$$

where ω is the forced frequency and P is the amplitude of the force. The solution of (7.17) consists of a *complementary* solution for the

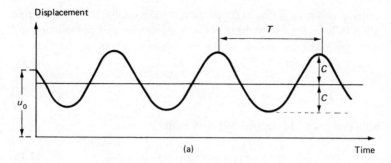

(a)

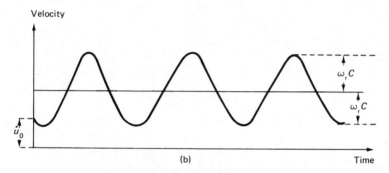

(b)

Figure 7.2   Plots of (a) displacement, (b) velocity

Figure 7.3   Forced vibrations of a spring–mass system

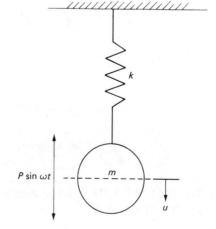

homogeneous equation and a *particular* solution for the F loading. In what follows we always neglect the transient part or complementary solution and concentrate on the particular solution. A trivial particular solution for (7.17) is:

$$u = U \sin \omega t \qquad (7.18)$$

Substituting (7.18) into (7.17) we obtain:

$$(- m\omega^2 U + kU) \sin \omega t = P \sin \omega t \qquad (7.19)$$

Hence:

$$U = \frac{P}{k - m\omega^2} \qquad (7.20)$$

or

$$u = \frac{P}{k - m\omega^2} \sin \omega t \qquad (7.21)$$

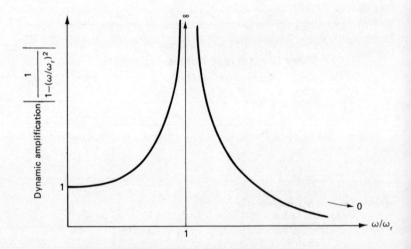

Figure 7.4   Response of spring–mass system

We could add (7.21) to (7.11) to have the general solution, but as we are not interested in the initial state let us assume that $u_0 = \dot{u}_0 = 0$

and investigate the behaviour of equation (7.21), which can be written as:

$$u = \frac{P}{k} \frac{1}{1 - (\omega/\omega_r)^2} \sin \omega t$$

$$= u_s \left| \frac{1}{1 - (\omega/\omega_r)^2} \right| \sin \omega t$$

(7.22)

where u_s is the static deflection. The term between brackets is called the 'dynamic amplification' of the system and can be plotted as a function of ω, as shown in Figure 7.4.

Note that for $\omega = \omega_r$ the amplitude of vibration tends to infinity. This value ω_r is called the *resonance* frequency for the system.

7.2 FORCED VIBRATIONS OF A DAMPED SYSTEM

In practice the amplitude is bounded, owing to the damping of the system. If we consider the case of *viscous* damping represented by the dashpot of Figure 7.5, where the motion is resisted by the viscosity of the fluid, a new force F_v can be added to equation (7.4), i.e.

$$F_a = -ku + F_v$$

(7.23)

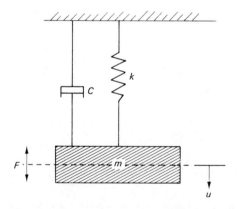

Figure 7.5   Dashpot–spring–mass system

This force acts in the same direction as the spring force and is equal to $-c\dot{u}$. Hence:

$$m\ddot{u} + ku + c\dot{u} = 0 \tag{7.24}$$

For the case of forced response we can write:

$$F = P \sin \omega t \tag{7.25}$$

where P is the amplitude of the exciting force and ω the forced frequency. Hence:

$$m\ddot{u} + c\dot{u} + ku = P \sin \omega t \tag{7.26}$$

For the particular integral we can try the following solutions:

$$u = A \cos \omega t + B \sin \omega t \tag{7.27}$$

which can also be expressed as:

$$u = U \sin (\omega t - \alpha) \tag{7.28}$$

which implies that $A = -U \sin \alpha$, $B = U \cos \alpha$. Substitution of this solution into (7.26) gives:

$$U \sin \omega t \left\{ c\omega \sin \alpha + k \cos \alpha - \frac{P}{U} - \omega^2 m \cos \alpha \right\} +$$

$$+ U \cos \omega t \left\{ c\omega \cos \alpha - k \sin \alpha + \omega^2 m \sin \alpha \right\} = 0 \tag{7.29}$$

The first term between brackets gives:

$$\frac{P}{U} = c\omega \sin \alpha + k \cos \alpha - \omega^2 m \cos \alpha$$

$$= \left\{ c\omega \tan \alpha + k - \omega^2 m \right\} \cos \alpha \tag{7.30}$$

and the second term gives:

$$\tan \alpha = \frac{c\omega}{k - \omega^2 m} \tag{7.31}$$

Thus the amplitude ratio is:

$$\frac{U}{P} = \frac{1}{\left[c\omega^2 + (k - \omega^2 m)^2 \right]^{\frac{1}{2}}} \tag{7.32}$$

It is useful to represent the exciting force (7.25) and the solution (7.28) in vector form (Figure 7.6). The angle α represents the difference

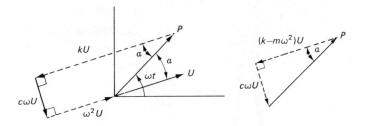

Figure 7.6   *Vector representation of forced damped vibrations*

in phase between the applied force F and the response u. This is produced by the damping term.

The equilibrium equation (7.26) could now be written as:

$$\underbrace{m\omega^2 U \sin(\omega t - \alpha)}_{\text{inertia}} - \underbrace{c\omega U \sin\left(\omega t - \alpha + \frac{\pi}{2}\right)}_{\text{damping}} - \underbrace{kU \sin(\omega t - \alpha)}_{\text{stiffness}} +$$

$$+ \underbrace{P\sin\omega t}_{\text{applied force}} = 0$$

We see that the damping force leads the displacement by 90°. This is because it is in the opposite direction to the velocity. The inertia force instead is in phase with the displacement. The vector interpretation of equations (7.31) and (7.32) is now quite evident. We can investigate these expressions further by writing the second in the form of equation (7.22), i.e.

$$u = \left(\frac{P}{k}\right) \frac{1}{\{[1 - (\omega/\omega_r)^2]^2 + 2(\omega/\omega_r)\}^{\frac{1}{2}}} \qquad (7.33)$$

where $\omega_r = k/m$ and $\gamma = c/2m\omega_r = $ damping factor. We can now plot (u/u_s), which is called the magnification factor and is a function only of damping and frequency (Figure 7.7). Note that $\tan \alpha$ can also be written as:

$$\tan \alpha = \frac{2\gamma(\omega/\omega_r)}{1 - (\omega/\omega_r)^2} \qquad (7.34)$$

The phase angle α is also plotted in Figure 7.7.

It is interesting to note that in the region $\omega/\omega_r \ll 1$ the angle α is small and the situation shown in Figure 7.6 applies. At resonance

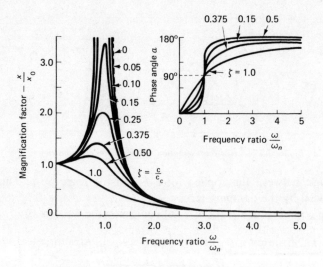

Figure 7.7   Magnification factor and phase angle as function of frequency

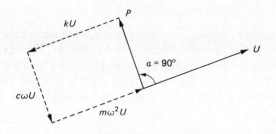

Figure 7.8   Resonance vector diagram

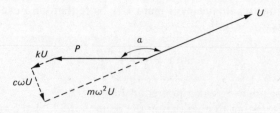

Figure 7.9   Inertia-dominant behaviour

$\omega = \omega_r$ the vector representation of the vibration can be seen in Figure 7.8, where now $\alpha = 90°$. The amplitude at resonance is:

$$U = \frac{P}{c\omega_r} = \frac{u_s}{2\gamma}$$

When $\omega/\omega_r \gg 1$ the angle α tends to $180°$ and the force P is used mainly to overcome the large inertia of the system (Figure 7.9).

7.3 COMPLEX RESPONSE METHOD

The use of complex algebra simplifies the forced vibration response of a damped system and it is important in order to find the random vibration response of the system. We can assume that the two functions, for the impressed form F and the displacement u, can be represented as:

$$F = P \exp(i\omega t) \qquad (7.35)$$

$$u = U^* \exp[i(\omega t - \alpha)] \qquad (7.36)$$

Note that the U response lags behind the applied force by the α angle.

For the case of a force such as the one given in equation (7.25) we have:

$$F = P \operatorname{Im}\{\exp(i\omega t)\}$$
$$\qquad (7.37)$$
and $\qquad u = U^* \operatorname{Im}\{\exp[i(\omega t - \alpha)]\}$

where $\operatorname{Im}\{\ \}$ means the imaginary part of the complex number. If the impressed force F were a cosine we would take:

$$\qquad (7.38)$$
$$F = P \operatorname{Re}\{\exp(i\omega t)\}$$
$$u = U^* \operatorname{Re}\{\exp[i(\omega t - \alpha)]\}$$

where $\operatorname{Re}\{\ \}$ means the real part of the complex number. In general we can work with equations (7.35) and (7.36), defining:

$$F = P\exp(i\omega t)$$
$$\qquad (7.39)$$
$$u = U \exp(i\omega t)$$

where $U = U^* \exp(-i\alpha)$.

Let us apply this complex analysis to equation (7.26), written now as:

$$m\ddot{u} + c\dot{u} + ku = P\exp(i\omega t) \qquad (7.40)$$

If the solution is of the form (7.39) we obtain:

$$(-m\omega^2 + i\omega c + k)U = P \qquad (7.41)$$

The complex response U is then:

$$U = \frac{P}{(k - m\omega^2) + ic\omega} \qquad (7.42)$$

Since $U = U^*\exp(-i\alpha)$ we can write:

$$U = U^*\exp(-i\alpha) = \frac{P}{[(k - m\omega^2)^2 + (\omega c)^2]^{\frac{1}{2}}}\exp(-i\alpha) \quad (7.43)$$

and the phase angle is, as before:

$$\tan\alpha = \frac{c}{k - m\omega^2} \qquad (7.44)$$

It is important to note that P and U do not need to be real, and more generally can be taken as complex. The function F then becomes similar to a term of the complex Fourier series seen in Chapter 2.

Equation (7.42) is usually written as:

$$U = \frac{P}{(k - m\omega^2) + ic\omega} = H(\omega)P \qquad (7.45)$$

where $H(\omega)$ is the complex frequency response function. Note that the complex frequency response can be written as:

$$H(\omega) = \frac{1}{k\left[1 - (\omega/\omega_r)^2 + 2i\gamma(\omega/\omega_r)\right]} \qquad (7.46)$$

The result is the same as the one we obtain by applying the Fourier transform to equation (7.41), i.e.

$$(-\omega^2 m + ic\omega + k)\overline{U} = \overline{P} \qquad (7.47)$$

or $$\overline{Z}(\omega)\overline{U} = \overline{P}$$

Solving this system we obtain:

$$\overline{U} = \overline{H}(\omega)\overline{P} \qquad (7.48)$$

with $\bar{H}(\omega) = \bar{Z}(\omega)^{-1}$. Applying the conjugate of the Fourier transform, we have:

$$(-\omega^2 m - ic\omega + k)\hat{U} = \hat{P} \qquad (7.49)$$

Hence for this case:

$$\hat{Z}(\omega) = (-\omega^2 m - ic\omega + k)$$
$$\hat{H}(\omega) = \hat{Z}(\omega)^{-1} \qquad (7.50)$$

$$\therefore \qquad \hat{U}(\omega) = \hat{H}(\omega)\hat{P} \qquad (7.51)$$

7.4 SPECTRAL DENSITY APPROACH

Having determined the Fourier transform and its conjugate, it is now possible to define the spectral densities of forces and displacements. We start by multiplying both members of equations (7.48) and (7.51) and dividing them by the period T. This gives:

$$\frac{1}{T}\,\bar{U}\hat{U} = \bar{H}(\omega)\,\frac{1}{T}\,\bar{P}\hat{P}\hat{H}(\omega) \qquad (7.52)$$

Remember that when $T \to \infty$ we obtain the spectral density (see section 2.2), i.e.

$$\lim_{T \to \infty}\left\{\frac{1}{T}\,|\bar{U}|^2\right\} = H(\omega)\lim_{T \to \infty}\left\{\frac{1}{T}\,|\bar{P}|^2\right\}\hat{H}(\omega) \qquad (7.53)$$

Therefore (7.52) becomes:

$$S_{uu}(\omega) = \bar{H}(\omega)\,S_{pp}(\omega)\,\hat{H}(\omega) \qquad (7.54)$$

or

$$S_{uu}(\omega) = |\bar{H}(\omega)|^2\,S_{pp}(\omega) \qquad (7.55)$$

This expression relates the spectral density of the forces to the spectral density of the response or displacement. We can now obtain the variance of the displacements:

$$\langle u^2 \rangle = \sigma_{uu}^2 = \int_0^\infty S_{uu}\,d\omega \qquad (7.56)$$

7.5 APPROXIMATE SOLUTIONS

The analysis of structural systems usually requires the use of an approximate method, 'exact' solutions being limited to very simple structure configurations. Two approximate methods of analysis are the finite-element method and the boundary-element one, which in contrast to finite elements discretises only the external surface of the continuum. Both methods are based on weighted-residual principles. It is important to understand how these principles can be applied in practice. In what follows we illustrate them for a simple column such as the one shown in Figure 7.10. The system will be reduced to a one-degree-of-freedom system after a series of simplifications.

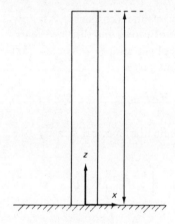

Figure 7.10   Simple column

The equilibrium equation for a beam element is the following fourth-order equation:

$$A\rho\ddot{u} + EI\,\frac{\mathrm{d}^4 u}{\mathrm{d}z^4} = p(x) \qquad (7.57)$$

where $p(x)$ are the applied forces along the beam, E is the modulus of elasticity and I the moment of inertia; u are the transverse displacements and A is the cross-sectional area. The boundary conditions are of two types:

● *Essential or displacement conditions* on the S_1 part of the boundary, of the type:

$$u = \bar{u} \quad \text{or} \quad \theta = \bar{\theta} \qquad (7.58)$$

where the bar indicates a known value of displacement u or rotation θ ($\theta = du/dz$).

● *Natural or force conditions* on the S_2 part of the boundary, which are moment M and shear Q:

$$\bar{M} = EI\,\frac{d^2u}{dz^2}, \quad \bar{Q} = -EI\,\frac{d^3u}{dz^3} \tag{7.59}$$

The bar denotes the applied (known) forces. Those without it are the internal components. Note that $S = S_1 + S_2$.

Initial conditions are not needed as we are only considering the steady-state solution.

One way of finding an approximate solution for these equations is by weighting equation (7.57) and the boundary conditions (7.58) and (7.59) in the following way:

$$\int_0^l \left(EI\,\frac{d^4u}{dz^2} + \rho A\ddot{u} - p\right)W\,dz$$

$$= \left[(\bar{M} - M)\,\frac{dW}{dx}\right]_{S_2} + \left[(\bar{Q} - Q)W\right]_{S_2} \tag{7.60}$$

where W are weighting functions that are assumed to satisfy the essential boundary conditions, i.e. W and dW/dz are identically zero on S_1. We assume that the shapes of the W functions are the same as the u functions we take as approximate solutions; this leads to the following form of the principle of virtual displacements:

$$\int_0^l \left(EI\,\frac{d^4u}{dz^2} + \rho A\ddot{u} - p\right)\delta u\,dz$$

$$= \left[(\bar{M} - M)\,\frac{d\delta u}{dz}\right]_{S_2} + \left[(\bar{Q} - Q)\delta u\right]_{S_2} \tag{7.61}$$

Integrating equation (7.61) by parts twice, we obtain the best-known expression for virtual displacements, i.e.

$$\int_0^l \left\{EI\left(\frac{d^2u}{dz^2}\right)\left(\frac{d^2\delta u}{dz^2}\right) + \rho A\ddot{u}\right\}dz$$

$$= \int_0^l p\,\delta u\,dz + \left[\bar{M}\,\frac{d\delta u}{dx}\right]_{S_2} + \left[\bar{Q}\,\delta u\right]_{S_2} \tag{7.62}$$

Consider now the beam shown in Figure 7.10, with boundary conditions:

$$u = \theta = 0 \quad \text{at } z = 0 \tag{7.63}$$

and

$$\overline{M} = \overline{Q} = 0 \quad \text{at } z = l \tag{7.64}$$

The principle of virtual displacements, equation (7.62), then becomes:

$$\int_0^l \left\{ EI \left(\frac{\mathrm{d}^2 u}{\mathrm{d}z^2} \right) \left(\frac{\mathrm{d}^2 \delta u}{\mathrm{d}z^2} \right) + \rho A \ddot{u} \right\} \mathrm{d}z = \int_0^l p \, \delta u \, \mathrm{d}z \tag{7.65}$$

with the function for u satisfying the essential boundary conditions (7.63) at $z = 0$, that is the part of the boundary S_2.

If we now assume an approximate function for u, such that:

$$u_A = u \, g(z)$$
$$\tag{7.66}$$
then similarly $\qquad \ddot{u}_A = \ddot{u} \, g(z)$

where u represents the horizontal displacement at the top of the column and $g(z)$ is a 'shape' function, i.e. it represents the shape of the column. Substituting (7.66) into (7.65) we find:

$$\delta u \left\{ \int_0^l EI \left(\frac{\mathrm{d}^2 g}{\mathrm{d}z^2} \right)^2 \mathrm{d}z \, u + \int_0^l \rho A (g)^2 \, \mathrm{d}z \, \ddot{u} \right\} = \delta u \left\{ \int_0^l pg \, \mathrm{d}z \right\} \tag{7.67}$$

As δu is arbitrary we can simply write:

$$Ku + M\ddot{u} = F \tag{7.68}$$

where

$$K = \int_0^l EI \left(\frac{\mathrm{d}^2 g}{\mathrm{d}z^2} \right) \mathrm{d}z \quad M = \int_0^l \rho A (g)^2 \, \mathrm{d}z \quad F = \int_0^l pg \, \mathrm{d}z \tag{7.69}$$

K, M and F are the equivalent stiffness, mass and force coefficients for the one-degree-of-freedom system.

We can similarly include the damping term into the equation. This will be illustrated in what follows. The more general equilibrium equation will then be:

$$Ku + C\dot{u} + M\ddot{u} = F \tag{7.70}$$

where

$$C = \int_0^l c(g)^2 \, dz \tag{7.71}$$

7.6 APPLICATION OF ONE-DEGREE-OF-FREEDOM SYSTEM ANALYSIS

Consider a concrete offshore structure, which can be idealised as shown in Figure 7.11. The complexity of a typical offshore structure is such that this idealisation is suspect. However, the response of these structures tends to occur predominantly in the first mode, which indicates that a one-degree-of-freedom model may be useful as a preliminary design tool. In addition the analysis gives us an insight into the way in which the more complex cases described in Chapter 8 can be solved.

Note that the column is of length l in a sea of depth d. The mass of

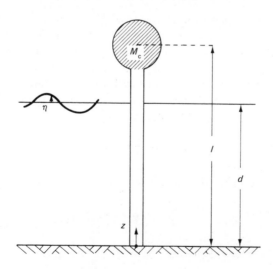

Figure 7.11   One-degree-of-freedom idealisation

the platform, m_c, is assumed to be concentrated at $z = l$. The stiffness and area are constant.

For the idealisation shown in Figure 7.11 we have the following equivalent values:

EI equivalent stiffness of the column, N m^2

A_c equivalent (concrete) area of the column, m^2

In addition we define:

A volume of water displaced per unit length, i.e. the cross-sectional area, m^2

m_c mass of the platform, kg

ρ density of the water, kg/m^3

ρ_c density of the concrete, kg/m^3

We have the following drag and inertia coefficients:

$$C_D = c_d \rho D/2 \ \ [\text{kg/m}^2], \quad C_M = c_m \rho A \ \ [\text{kg/m}],$$
$$C_A = \rho A \ \ [\text{kg/m}]$$

where c_d is the drag coefficient (1.0 for cylinders) and c_m is the inertia coefficient (also 1.0 for cylinders). The equilibrium equation is (7.70):

$$M\ddot{u} + C\dot{u} + Ku = F(t) \tag{7.72}$$

where u is the displacement at the top of the column ($x = l$). The term C includes the structural and hydrodynamic damping. M is obtained by addition of the mass of the column, the mass of the platform and the hydrodynamic mass. Note that the C_A term does not enter into M because C_A only affects the water particle accelerations.

If the shape of the column is assumed to be $g(\bar{z})$, where $\bar{z} = z/l$, the M term in equation (7.72) becomes:

$$M = l\rho_c A_c \int_0^1 [g(\bar{z})]^2 \, d\bar{z} + C_M l \int_0^{d/l} [g(\bar{z})]^2 \, d\bar{z} + m_c \tag{7.73}$$

The inertia term for the column is:

$$K = \frac{EI}{l^3} \int_0^1 \left(\frac{\partial^2 g(\bar{z})}{\partial \bar{z}^2} \right)^2 d\bar{z} \tag{7.74}$$

The natural frequency of the system is:

$$\omega_r = \sqrt{(K/M)} \tag{7.75}$$

In order to write the C term (where C is formed by the addition of the structural plus hydrodynamic damping: $C = C_s + C_H$) in its usual form $C = 2M\gamma\omega_r$, let us consider the hydrodynamic damping term. The drag coefficient in Morison's formula was written as:

$$\overline{C}_D = C_D\sqrt{(8/\pi)}\sigma_{v_x} \tag{7.76}$$

and is multiplied by $(v_x - \dot{u})$, where the $\dot{u}$ contribution can be passed to the left-hand side and combined with C_s.

The hydrodynamic damping contribution for the column is:

$$C_H\dot{u} = \left\{C_D\sqrt{(8/\pi)}\int_0^{d/l}\sigma_{v_x}[g(\bar{z})]^2\,d\bar{z}\right\}\dot{u} \tag{7.77}$$

One needs first to compute the variance of the water particle velocity, σ_{v_x} which is a function of the velocity spectrum, i.e.

$$\sigma_{v_x}^2 = \int_0^\infty S_{v_x v_x}(\omega)\,d\omega = \int_0^\infty \omega^2\frac{\cosh^2\kappa z}{\sinh^2\kappa d}S_{\eta\eta}(\omega)\,d\omega \tag{7.78}$$

Hence the deviation is:

$$\sigma_{v_x} = \sqrt{\sigma_{v_x}}\quad[\text{m}^2/\text{s}] \tag{7.79}$$

Once the term C_H has been computed we can write the percentage of critical damping as:

$$\gamma = \gamma_s + \frac{C_H}{2M\omega_r} \tag{7.80}$$

where γ_s is the structural damping contribution. The equation of motion can now be written as:

$$M\ddot{u} + 2\gamma M\omega_r\dot{u} + \omega_r^2 Mu = F(t) \tag{7.81}$$

The $F(t)$ term is computed from the contribution of the $\dot{v}$, v terms in Morison's equation, i.e.

$$P_w = (C_M + C_A)\dot{v} + C_D\sqrt{(8/\pi)}\sigma_{v_x}v$$

$$= (C_M + C_A)\omega^2\frac{\cosh\kappa z}{\sinh\kappa d}\eta + C_D\sqrt{\left(\frac{8}{\pi}\right)}\sigma_{v_x}\omega\frac{\cosh\kappa z}{\sinh\kappa d}\eta' \tag{7.82}$$

where $\eta = a\cos\omega t$ and $\eta' = a\sin\omega t$. (Note that the column is now taken to be at $x = 0$.)

The generalised force for the column can now be written as:

$$F(t) = \int_0^d P g(\bar{z}) \, dz$$

or

$$F(t) = (C_M + C_A) \frac{\omega^2}{\sinh \kappa d} \eta \int_0^d \cosh \kappa z \, g(\bar{z}) \, dz +$$

$$+ C_D \sqrt{\left(\frac{8}{\pi}\right)} \frac{\omega}{\sinh \kappa d} \eta' \int_0^d \cosh \kappa z \, g(\bar{z}) \, \sigma_{v_x} \, dz \quad (7.83)$$

The spectral density function for this force is:

$$S_{FF}(\omega) = \left[\frac{(C_M + C_A)^2}{\sinh^2 \kappa d} \omega^4 \left\{ \int_0^d \cosh \kappa z \, g(\bar{z}) \, dz \right\}^2 + \right.$$

$$\left. + C_D \frac{8}{\pi} \frac{\omega^2}{\sinh^2 \kappa d} \left\{ \int_0^d \cosh \kappa z \, \sigma_{v_x} \, g(\bar{z}) \, dz \right\}^2 \right] S_\eta \quad (7.84)$$

Note that in this case the cross spectral density terms relating v_x and $\dot{v}_x$ disappear, owing to their different phases.

The transfer function for displacements is obtained by substituting $u = \bar{U} \exp(i\omega t)$, $F = \bar{F} \exp(i\omega t)$ into the equation of motion. This gives:

$$M(-\omega^2 + 2i\gamma\omega\omega_r + \omega_r^2) \bar{U} = \bar{F} \quad (7.85)$$

Hence:

$$H(\omega)\bar{F}(\omega) = \bar{U}(\omega)$$

with $H(\omega) = \left[(-\omega^2 + 2i\gamma\omega\omega_r + \omega_r^2)M \right]^{-1}$

The following relationship applies for the spectral densities:

$$|H(\omega)|^2 S_{FF}(\omega) = S_{\bar{U}\bar{U}}(\omega) \quad (7.86)$$

where

$$|H(\omega)|^2 = \frac{1}{M^2 \omega_r^4 \left\{ \left[1 - \left(\frac{\omega}{\omega_r}\right)^2 \right]^2 + \left(2\gamma \frac{\omega}{\omega_r}\right)^2 \right\}} = \frac{|H'(\omega)|^2}{M^2 \omega_r^4}$$

$$(7.87)$$

The spectral density of the force is:

$$S_{FF}(\omega) = |A(\omega)|^2 S_{\eta\eta} \qquad (7.88)$$

where $|A(\omega)|^2$ is the force transfer function defined in equation (7.84). Once $S_{\bar{U}\bar{U}}(\omega)$ is known its variance can be calculated:

$$\sigma_{\bar{U}}^2 = \int_0^\infty S_{\bar{U}\bar{U}}(\omega)\,d\omega \qquad (7.89)$$

As one is working with a Gaussian process with zero mean, it is possible to calculate the probability of $\bar{U}$ being within a certain value $\pm \lambda \sigma_U$; for instance, for $\lambda = 3$ the probability is 99.7 per cent. Alternatively, knowing that the peaks of a narrow-band Gaussian process have a Rayleigh distribution, i.e.

$$p(\bar{U}) = \frac{\bar{U}}{\sigma_{\bar{U}}^2} \exp\left(\frac{-\bar{U}}{2\sigma_{\bar{U}}^2}\right) \qquad (7.90)$$

one can compute the most probable maximum deflection (or stress) for a given storm.

The expected maximum value of the response can be approximated by:

$$\langle |\bar{U}|_{\max} \rangle = \sigma_{\bar{U}} \left\{ \sqrt{[2\ln(T/T_{\mathrm{m}})]} \right\} \qquad (7.91)$$

where T is the duration of the storm

T_{m} is the mean period, given by:

$$T_{\mathrm{m}} = 2\pi \left\{ \frac{\displaystyle\int_0^\infty S_{\bar{U}\bar{U}}\,d\omega}{\displaystyle\int_0^\infty \omega^2\, S_{\bar{U}\bar{U}}\,d\omega} \right\}^{\frac{1}{2}} \qquad (7.92)$$

In addition to $\sigma_{\bar{U}}$ we can calculate the variance of any other quantity such as stresses or moments. Assume, for instance, that the bending moment at the base of the column is related to the displacement at the top by a function B such that:

$$M = B\bar{U} \qquad (7.93)$$

The spectral density for this moment is now:

$$S_{MM}(\omega) = B^2 S_{\bar{U}\bar{U}}(\omega) \qquad (7.94)$$

Example 7.1

Assume we have an offshore structure that can be approximated to the one-degree-of-freedom structure shown in Figure 7.11 with the following characteristics:

$$d = 75\,\text{m} \qquad\qquad l = 100\,\text{m}$$

$$m_c = 2 \times 10^6\,\text{kg} \qquad EI = 2250 \times 10^9\,\text{N m}^2 \qquad \text{(a)}$$

$A_c = 29\,\text{m}^2$ (cross-sectional area of concrete)

$A = 78\,\text{m}^3/\text{m}$ (total volume of water displaced per unit length)

$\rho = 10^3\,\text{kg/m}^3$ (density of water)

$\rho_c = 2.5 \times 10^3\,\text{kg/m}^3$ (density of concrete)

$D = 10\,\text{m}$ (diameter of the column)

The drag and inertia coefficients for the equivalent column are:

$$C_D = \frac{c_d \rho D}{2} = 5000\,\text{kg/m}^2$$

$$C_M = c_m \rho A = 78\,000\,\text{kg/m} \qquad\qquad \text{(b)}$$

$$C_A = \rho A \quad = 78\,000\,\text{kg/m}$$

($c_m = c_d = 1$ for this case).

The wave spectrum used is the one given by Pierson and Moskowitz for a wind velocity $W = 20\,\text{m/s}$. The deflected shape of the structure will be approximated by $g(\bar{z}) = z^2$, where $\bar{z} = z/l$. Hence the mass of the system can be written:

$$M = l\rho_c A_c \int_0^1 \bar{z}^4\,d\bar{z} + C_M l \int_0^{3/4} \bar{z}^4\,d\bar{z} + m_c = 3860 \times 10^3\,\text{kg} \quad \text{(c)}$$

and the stiffness is:

$$K = \frac{EI}{l^3} \int_0^1 \frac{\partial^2 g}{\partial \bar{z}^2}\,d\bar{z} = 9000 \times 10^3\,\text{N/m} \qquad \text{(d)}$$

One can now find the natural frequency of the system, ω_r:

$$\omega_r^2 = \frac{K}{M} = \frac{9000 \times 10}{3860 \times 10} = 2.334$$

$$\omega_r = 1.54\,\text{radians/second} \qquad\qquad \text{(e)}$$

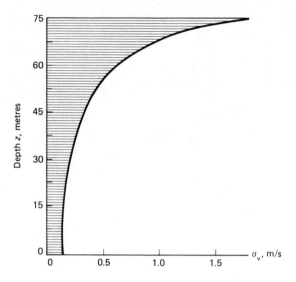

Figure 7.12   Variation of water particle velocity with depth

In order to calculate the damping constant γ one needs to compute the hydrodynamic damping constant C_H, having first found the variance σ_{v_x} at different heights. The variation of σ_{v_x} for the spectrum under consideration is shown in Figure 7.12 and was obtained by integrating numerically equation (7.78). Next one calculates $C_H \dot{u}$ using equation (7.77), which gives:

$$C_H \dot{u} = 113\,000\,\dot{u} \qquad\qquad (f)$$

The damping constant can now be found, i.e.

$$\gamma = \gamma_s + \frac{113\,000}{118\,900} \times 10^{-2}$$

$$\simeq \gamma_s + 0.01 \qquad\qquad (g)$$

For the structural damping the value $\gamma_s = 0.05$ was taken. Hence $\gamma = 0.06$ for this case. The next step is to evaluate using numerical integration the force spectra given by equation (7.84). They are shown in Figure 7.13, where the drag and inertia contributions can be seen. The transfer function $H(\omega)$ for a one-degree-of-freedom system can be computed using formulae (7.87), and the results are plotted in Figure 7.14. Next the values of the transfer function are multiplied by

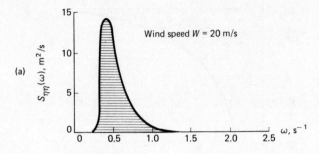

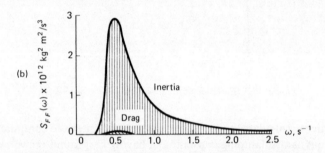

Figure 7.13 Spectral density of (a) waves, (b) generalised force F

the spectral density S_{FF} to obtain the response spectrum, i.e.

$$S_{\bar{U}\bar{U}} = |H(\omega)|^2 S_{FF} = |H'(\omega)|^2 \frac{1}{M^2 \omega_r^4} S_{FF} \qquad \text{(h)}$$

which is shown in Figure 7.14. Integrating this spectrum, the variance of the generalised displacement can be obtained, i.e.

$$\sigma_{\bar{U}}^2 = \int_0^\infty S_{\bar{U}\bar{U}}(\omega)\, d\omega \qquad \text{(i)}$$

Numerical integration of equation (i) gives:

$$\sigma_{\bar{U}}^2 = 0.092\,\text{m}^2$$

$$\therefore \qquad \sigma_{\bar{U}} = 0.304\,\text{m}$$

The probability of the $\bar{U}$ value being within $\pm 3\sigma_{\bar{U}} = \pm 0.912\,\text{m}$ is

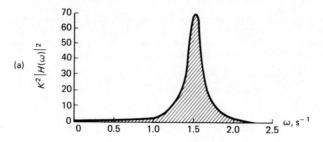

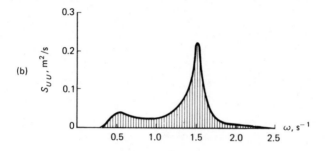

Figure 7.14 (a) Transfer function, (b) spectral density of response

99.7 per cent. The moment can also be obtained, i.e.

$$\sigma_M^2 = (EI/l^2)^2 \, \sigma_{\bar{U}}^2$$

$$\sigma_M = 1.35 \times 10^5 \, \text{N m} \tag{j}$$

$$\pm 3\sigma_M = \pm 4.05 \times 10^5 \, \text{N m}$$

Alternatively we could have calculated the multiplier of $\sigma_{\bar{U}}$ using equation (7.91), i.e.

$$\langle \, |\bar{U}|_{\text{max}} \, \rangle = \sigma_{\bar{U}} \{ \sqrt{[2\ln(T/T_{\text{m}})]} \} \tag{k}$$

Bibliography

Brebbia, C. A., et al., Vibrations of engineering structures, Southampton University Press (1974)

Brebbia, C. A., and Connor, J. J., Fundamentals of finite element techniques for structural engineers, Butterworths (1973)

Dym, C. L., and Shames, I. H., *Solid mechanics: a variational approach*, McGraw-Hill (1973)
Thomson, W. T., *Vibration theory and applications*, Allen and Unwin (1966)
Warburton, G. B., *The dynamical behaviour of structures*, Pergamon Press (1964)

8 Discretisation for Multi-degree-of-freedom Systems

8.1 INTRODUCTION

Offshore structures can be represented as multi-degree-of-freedom systems, consisting of combinations of elements such as beams, columns, plates, etc. The properties of each element are analysed separately and written in matrix form. Combination of all the elements produces a global set of equations, which is also written in matrix form.

A method of obtaining these element matrices is to use the finite element displacement technique. This consists of taking the displacement measures at discrete points in the body as unknowns and defining the displacement field in terms of these discrete variables, much as was done for the one-degree-of-freedom column in Chapter 7. Once the discrete displacements are known, the strains are computed from the strain–displacement relations and then the stresses are determined from the stress–strain relations. The method is an application of the weighted residual technique, but in contrast with more classical methods the displacement expansions are applied only on a part or 'element' of the domain.

In the displacement method, the application of weighted residual techniques results in a set of simultaneous algebraic equations for the unknown displacements. Because of the large number of variables, the analysis is most conveniently formulated in terms of matrix algebra. The stiffness and mass matrices for the structure are obtained by superimposing the contribution of the element stiffness and mass matrices at each node. The system load vector is generated in a similar way, i.e. by superimposing the element force vectors. The displacement boundary conditions are then enforced. These steps result in a set of algebraic equations relating the displacement measures.

The various steps of the method are:

● Discretisation of the body, i.e. selection of elements interconnected at certain nodal points.

● Evaluation of the element stiffness, mass and force matrices.

● Assemblage of the mass and force matrices for the system of elements and nodes (system equations) and introduction of the displacement boundary conditions.

● Solution of the resulting system equations and calculation of strains and stresses based on the nodal displacements.

8.2 FORMULATION FOR A BEAM ELEMENT

In the finite element method we consider the body to be divided into volume elements having finite dimensions, and we select certain points on the interior and exterior boundary surfaces. The volume elements are referred to as 'finite' elements since their dimensions are finite; the boundary points are called nodal points or nodes. We number the elements and nodes and specify the element–node connectivity by listing, for each element, the nodes associated with that element. A typical discretisation for a frame is shown in Figure 8.1. The element connectivity table is shown in the figure. It is irrelevant which is the starting node.

Next, we define nodal displacement quantities. The number and choice of displacement quantities is problem-dependent, but they have to be at least the displacements that satisfy the boundary conditions on S_1. For a beam element such as shown in Figure 8.2 they are the nodal displacements and their rotations.

Let us now consider a simple beam element such as the one shown in Figure 8.2. The dynamic equilibrium equation for this element is:

$$EI \frac{d^4 w}{dx^4} + A\rho \ddot{w} = p(x) \tag{8.1}$$

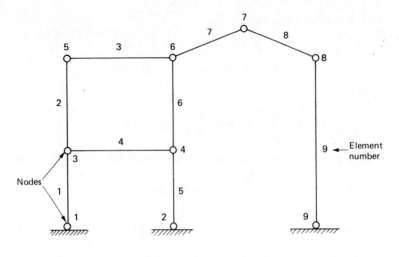

Element	Node	
	n_1	n_2
1	1	3
2	3	5
3	5	6
4	3	4
5	2	4
6	4	6
7	6	7
8	7	8
9	8	9

Figure 8.1   Typical frame structure

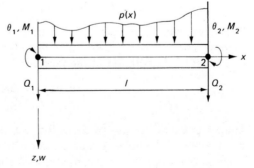

Figure 8.2   Beam element

with boundary conditions:

$$\text{(i)} \quad w = \bar{w} \quad \text{and} \quad \theta = \bar{\theta} \quad \text{on } S_1$$

$$\text{(ii)} \quad M = \bar{M} \quad \text{and} \quad Q = \bar{Q} \quad \text{on } S_2 \tag{8.2}$$

where

$$\theta = \frac{dw}{dx} \quad M = EI \frac{d^2 w}{dx^2} \quad Q = -EI \frac{d^3 w}{dx^3}$$

Equations (8.1) and (8.2) can be written in weighted residual form as:

$$\int_0^l \left[EI \frac{d^4 w}{dx^4} + A \rho \ddot{w} - \bar{p}(x) \right] \delta w \, dx$$

$$= \left| (M - \bar{M}) \, \delta \theta \right|_{S_2} + \left| (Q - \bar{Q}) \delta w \right|_{S_2} \tag{8.3}$$

The function w is assumed to satisfy the boundary conditions on S_1; hence $\delta w = \delta \theta = 0$ on S_1. Integrating (8.3) by parts twice, we obtain:

$$\int_0^l \left(EI \frac{d^2 w}{dx^2} \frac{d^2 \delta w}{dx^2} - p \, \delta w + A \rho \ddot{w} \, \delta w \right) dx = \left[\bar{M} \, \delta \theta + \bar{Q} \, \delta w \right]_{S_2} \tag{8.4}$$

Rearranging, we have:

$$\int_0^l \left(EI \frac{d^2 w}{dx^2} \frac{d^2 \delta w}{dx^2} + A \rho \ddot{w} \, \delta w \right) dx = \int_0^l p \, \delta w \, dx + \left[\bar{M} \, \delta \theta + \bar{Q} \, \delta w \right]_{S_2} \tag{8.5}$$

Equation (8.5) is the starting point for the formulation of the element matrices. The choice of nodal variables is v and θ, and it follows that the vector of element unknowns $\mathbf{U}_e$ is at least 1×4:

$$\mathbf{U}_e = \{ w_1 \, \theta_1 \, w_2 \, \theta_2 \ldots \} \tag{8.6}$$

This discussion is restricted to negligible transverse shear deformation. The extension strain varies linearly through the depth, and θ is equal to the rotation of the tangent.

$$\varepsilon = -z \frac{d^2 w}{dx^2} \quad \theta = \frac{dw}{dx} \tag{8.7}$$

Since ε involves the second derivative, the displacement expansion must contain a complete quadratic in order to be able to represent rigid body motion and constant strain. We write:

$$w = \alpha_1 + \alpha_2 x + \alpha_3 x^2 + \alpha_4 x^3 = \begin{bmatrix} 1 & x & x^2 & x^3 \end{bmatrix} \begin{Bmatrix} \alpha_1 \\ \alpha_2 \\ \alpha_3 \\ \alpha_4 \end{Bmatrix} = \mathbf{A}\boldsymbol{\alpha} \quad (8.8)$$

The $\boldsymbol{\alpha}$ may be expressed as functions of the generalised displacements w_i. Thus:

$$\begin{Bmatrix} w_1 \\ \theta_1 \\ w_2 \\ \theta_2 \end{Bmatrix} = \begin{bmatrix} 1 & 0 & 0 & 0 \\ 0 & 1 & 0 & 0 \\ 1 & l & l^2 & l^3 \\ 0 & 1 & 2l & 3l^2 \end{bmatrix} \begin{Bmatrix} \alpha_1 \\ \alpha_2 \\ \alpha_3 \\ \alpha_4 \end{Bmatrix}$$

or $\qquad\qquad\qquad \mathbf{U}_e = \mathbf{C}\boldsymbol{\alpha}$

$$(8.9)$$

Next, we invert (8.9):

$$\boldsymbol{\alpha} = \mathbf{C}^{-1}\mathbf{U}_e = \begin{bmatrix} 1 & 0 & 0 & 0 \\ 0 & 1 & 0 & 0 \\ -\dfrac{3}{l^2} & -\dfrac{2}{l} & \dfrac{3}{l^2} & -\dfrac{1}{l} \\ \dfrac{2}{l^3} & \dfrac{1}{l^2} & -\dfrac{2}{l^3} & \dfrac{1}{l^2} \end{bmatrix} \begin{Bmatrix} w_1 \\ \theta_1 \\ w_2 \\ \theta_2 \end{Bmatrix} \quad (8.10)$$

Substituting $\boldsymbol{\alpha}$ in (8.8) we obtain the final result:

$$w = \mathbf{G}\mathbf{U}_e = g_1 w_1 + g_2 \theta_1 + g_3 w_2 + g_4 \theta_2 \quad (8.11)$$

where $g_1 = \left[1 - 3\left(\dfrac{x}{l}\right)^2 + 2\left(\dfrac{x}{l}\right)^3 \right] \quad g_2 = \left[x - 2\dfrac{x^2}{l} + \dfrac{x^3}{l^2} \right]$

$\qquad g_3 = \left[3\left(\dfrac{x}{l}\right)^2 - 2\left(\dfrac{x}{l}\right)^3 \right] \qquad g_4 = \left[-\dfrac{x^2}{l} + \dfrac{x^3}{l^2} \right]$

g_i are the interpolation functions. Similarly:

$$\ddot{w} = \mathbf{G}\ddot{\mathbf{U}}_e \quad (8.12)$$

The strain expansion is obtained by substituting for w in (8.7):

$$\varepsilon = -z\frac{d^2 w}{dx^2} = -z\frac{d^2}{dx^2}(\mathbf{G})\mathbf{U}_e = \mathbf{B}\mathbf{U}_e \quad (8.13)$$

$$\mathbf{B} = -z\left\{ -\frac{6}{l^2} + \frac{12x}{l^3}, \ -\frac{4}{l} + \frac{6x}{l^2}, \ \frac{6}{l^2} - \frac{12x}{l^3}, \ -\frac{2}{l} + \frac{6}{l^2} \right\}$$

We can now substitute formulae (8.11) to (8.13) into (8.5), obtaining:

$$\delta \mathbf{U}_e^T \left\{ \int_0^l EI \mathbf{B}^T \mathbf{B} \, dx \, \mathbf{U}_e + \int_0^l A\rho \mathbf{G}^T \mathbf{G} \, dx \, \ddot{\mathbf{U}}_e \right\} = \delta \mathbf{U}_e^T \left\{ \int_0^l \mathbf{G}^T \bar{p} \, dx \right\}$$
(8.14)

where we have not written the $\bar{M}$ and $\bar{Q}$ forces, as they are concentrated forces that can easily be applied at a later stage. Since the $\delta \mathbf{U}_e$ are arbitrary, equation (8.14) can be written:

$$\mathbf{K} \mathbf{U}_e + \mathbf{M} \ddot{\mathbf{U}}_e = \mathbf{F}$$
(8.15)

where $\mathbf{K} = \int_0^l EI \mathbf{B}^T \mathbf{B} \, dx$ the stiffness matrix

$\mathbf{M} = \int_0^l A\rho \mathbf{G}^T \mathbf{G} \, dx$ the mass matrix

$\mathbf{F} = \int_0^l \mathbf{G}^T p \, dx$ the consistent force vector

These can now be evaluated.

Stiffness matrix

$$\mathbf{K} = \int \mathbf{B}^T EI \mathbf{B} \, dx = \frac{EI}{l^3} \begin{bmatrix} 12 & 6l & -12 & 6l \\ & 4l^2 & -6l & 2l^2 \\ & & 12 & -6l \\ \text{sym.} & & & 4l^2 \end{bmatrix}$$
(8.16)

This result is the 'exact' stiffness matrix for a prismatic beam. The agreement is due to the expansion employed. One can readily show that the exact homogeneous solution of the governing equations for a prismatic element is a cubic polynomial.

Mass matrix

$$\mathbf{M} = \int A\rho \mathbf{G}^T \mathbf{G} \, dx = \frac{\rho Al}{420} \begin{bmatrix} 156 & 22l & 54 & -13l \\ & 4l^2 & 13l & -3l^2 \\ & & 156 & -22l \\ \text{sym.} & & & 4l^2 \end{bmatrix}$$
(8.17)

Consistent element nodal force matrix for the case of only distributed transverse loading $p(x)$. This is:

$$\mathbf{F} = \int_0^{} \mathbf{G}^T p(x) \, dx \qquad (8.18)$$

or

$$\begin{Bmatrix} p_{1w} \\ p_{1\theta} \\ p_{2w} \\ p_{2\theta} \end{Bmatrix} = l \int_0^l \begin{Bmatrix} p(1 - 3\bar{x}^2 + 2\bar{x}^3) \\ pl(1 - 2\bar{x}^2 + \bar{x}^3) \\ p(3\bar{x}^2 - 2\bar{x}^3) \\ pl(-\bar{x}^2 + \bar{x}^3) \end{Bmatrix} d\bar{x}$$

where $\bar{x} = x/l$. The form of $\mathbf{F}$ will depend on how $p(x)$ varies. If p is constant, we obtain after integration:

$$\mathbf{F} = \begin{Bmatrix} \frac{1}{2} \\ \frac{1}{4}l \\ \frac{1}{2} \\ -\frac{1}{4}l \end{Bmatrix} pl$$

Example 8.1

Consider now the beam represented by two elements of equal length (Figure 8.3). For element ① we have:

$$\mathbf{K}^{①}\mathbf{U}^{①} + \mathbf{M}^{①}\ddot{\mathbf{U}}^{①} = \mathbf{F}^{①} \qquad (a)$$

and for element ②:

$$\mathbf{K}^{②}\mathbf{U}^{②} + \mathbf{M}^{②}\ddot{\mathbf{U}}^{②} = \mathbf{F}^{②} \qquad (b)$$

These equations can be assembled by using the compatibility (local

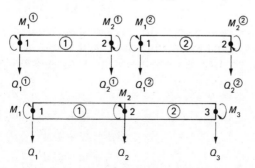

Figure 8.3 Beam composed of two elements

node 2 of beam ① is the same as node 1 of beam ②)) and equilibrium conditions:

$$Q_1 = Q_1^{①}, \quad Q_2 = Q_2^{①} + Q_1^{②}, \quad Q_2 = Q_2^{②}$$
$$M_1 = M_1^{①}, \quad M_2 = M_2^{①} + M_1^{②}, \quad M_2 = M_2^{②}$$
(c)

The above relationships imply that the coefficients of the mass and stiffness element matrices can now be written as:

Equilibrium equation that corresponds to

$$
\begin{matrix}
Q_1 & \rightarrow \\
M_1 & \rightarrow \\
Q_2 & \rightarrow \\
M_2 & \rightarrow \\
Q_3 & \rightarrow \\
M_3 & \rightarrow
\end{matrix}
\left[\cdots \right]
\text{ multiplied by }
\left\{
\begin{matrix}
\ddot{w}_1 \\
\ddot{\theta}_1 \\
\ddot{w}_2 \\
\ddot{\theta}_2 \\
\ddot{w}_3 \\
\ddot{\theta}_3
\end{matrix}
\right\}
\text{ or their acceleration}
$$
(d)

This gives:

$$
\frac{EI}{l^3}
\begin{bmatrix}
12 & 6l & -12 & 6l & 0 & 0 \\
 & 4l^2 & -6l & 2l^2 & 0 & 0 \\
 & & 24 & 0 & -12 & 6l \\
 & & & 8l^2 & -6l & 2l^2 \\
 & & & & 12 & -6l \\
\text{sym.} & & & & & 4l^2
\end{bmatrix}
\left\{
\begin{matrix}
w_1 \\
\theta_1 \\
\ddot{w}_2 \\
\ddot{\theta}_2 \\
\ddot{w}_3 \\
\ddot{\theta}_3
\end{matrix}
\right\} +
$$

$$
+ \frac{\rho A l}{420}
\begin{bmatrix}
156 & 22l & 54 & -13l & 0 & 0 \\
 & 4l^2 & 13 & -3l^2 & 0 & 0 \\
 & & 312 & 0 & 54 & -13l \\
 & & & 8l^2 & 13l & -3l^2 \\
 & & & & 156 & -22l \\
 & & & & & 4l^2
\end{bmatrix}
\left\{
\begin{matrix}
\ddot{w}_1 \\
\ddot{\theta}_1 \\
\ddot{w}_2 \\
\ddot{\theta}_2 \\
\ddot{w}_3 \\
\ddot{\theta}_3
\end{matrix}
\right\} =
\left\{
\begin{matrix}
Q_1 \\
M_1 \\
Q_2 \\
M_2 \\
Q_3 \\
M_3
\end{matrix}
\right\}
$$
(e)

Finally we can impose the displacement boundary conditions $w_1 = w_3 = 0$, and assume the right hand side of (e) is zero, i.e. the vibrations are free. We obtain the following system of equations:

$$
\begin{bmatrix}
(4-4\lambda)l^2 & (-6-12\lambda)l & (2+3\lambda)l^2 & 0 \\
 & (24-312\lambda)l & 0 & (6+13\lambda)l \\
 & & (8-8\lambda)l^2 & (2+3\lambda)l^2 \\
\text{sym.} & & & (4-4\lambda)l^2
\end{bmatrix}
\begin{Bmatrix}
\theta_1 \\
w_2 \\
\theta_2 \\
\theta_3
\end{Bmatrix}
=
\begin{Bmatrix}
0 \\
0 \\
0 \\
0
\end{Bmatrix}
$$
(f)

where $\lambda = \rho A l^4 \omega^2 / 420 \, EI$. The solution of the system of equations (f) gives the following eigenvalues:

$$\bar{\lambda} = \frac{\rho A L^4 \omega^2}{EI} = 98.18, \; 1920, \; 12130, \; 40320$$

($L = 2l$, total length of the beam). The exact solution for the beam gives:

$$\bar{\lambda} = 97.41, \; 1559, \; 7890, \; 24940$$

A more general beam element is the one shown in Figure 8.4, where the displacement functions for the element are:

$$
\begin{aligned}
w &= \alpha_1 + \alpha_2 x + \alpha_3 x^2 + \alpha_4 x^3 \\
v &= \alpha_5 + \alpha_6 x + \alpha_7 x^2 + \alpha_8 x^3 \\
u &= \alpha_9 + \alpha_{10} x \\
\gamma &= \alpha_{11} + \alpha_{12} x
\end{aligned}
\tag{8.19}
$$

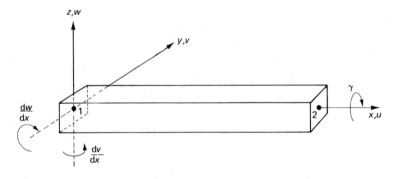

Figure 8.4 Three-dimensional beam element

γ is the twist angle. The element now has 12 parameters, which can be expressed in terms of the six nodal variables $(w, \partial w/\partial x, v, \partial v/\partial x, u, \gamma)$ at each nodal point. Note that the extension and twist are represented by a linear expansion while the two transverse deflections are cubic. The expression for the weighted residual statement is similar to the previous one, but will now include more terms. The final matrices are as follows.

Stiffness matrix

$$\frac{E}{l^3}\begin{bmatrix}
12I_z & 6lI_z & 0 & 0 & 0 & 0 & -12I_z & 6lI_z & 0 & 0 & 0 & 0\\
6lI_z & 4l^2I_z & 0 & 0 & 0 & 0 & -6lI_z & 2l^2I_z & 0 & 0 & 0 & 0\\
0 & 0 & 12I_y & 6lI_y & 0 & 0 & 0 & 0 & -12I_y & 6lI_y & 0 & 0\\
0 & 0 & 6lI_y & 4l^2I_y & 0 & 0 & 0 & 0 & -6lI_y & 2l^2I_y & 0 & 0\\
0 & 0 & 0 & 0 & l^2A & 0 & 0 & 0 & 0 & 0 & -l^2A & 0\\
0 & 0 & 0 & 0 & 0 & \dfrac{l^2GJ}{E} & 0 & 0 & 0 & 0 & 0 & -\dfrac{l^2GJ}{E}\\
-12I_z & -6lI_z & 0 & 0 & 0 & 0 & 12I_z & 6lI_z & 0 & 0 & 0 & 0\\
6lI_z & 2l^2I_z & 0 & 0 & 0 & 0 & 6lI_z & 4l^2I_z & 0 & 0 & 0 & 0\\
0 & 0 & -12I_y & -6lI_y & 0 & 0 & 0 & 0 & 12I_y & 6lI_y & 0 & 0\\
0 & 0 & 6lI_y & 2l^2I_y & 0 & 0 & 0 & 0 & 6lI_y & 4l^2I_y & 0 & 0\\
0 & 0 & 0 & 0 & -l^2A & 0 & 0 & 0 & 0 & 0 & l^2A & 0\\
0 & 0 & 0 & 0 & 0 & -\dfrac{l^2GJ}{E} & 0 & 0 & 0 & 0 & 0 & \dfrac{l^2GJ}{E}
\end{bmatrix} \qquad (8.20)$$

where E is Young's modulus

G is the shear modulus

A is the cross-sectional area

l is the element length

J is the polar 2nd moment of area

I_z, I_y are the 2nd moments of area about the relevant axes

Mass matrix

$$\frac{\rho A l}{420}\begin{bmatrix}
156 & & & & & & & & & & & \\
22l & 4l^2 & & & & & & & & & & \\
0 & 0 & 156 & & & & & & & & & \\
0 & 0 & 22l & 4l^2 & & & & & & & & \\
0 & 0 & 0 & 0 & 140 & & & & & & & \\
0 & 0 & 0 & 0 & 0 & \dfrac{140I_{\mathrm p}}{A} & & & & & & \\
54 & 13l & 0 & 0 & 0 & 0 & 156 & & & & & \\
-13l & -3l^2 & 0 & 0 & 0 & 0 & 22l & 4l^2 & & & & \\
0 & 0 & 54 & -13l & 0 & 0 & 0 & 0 & 156 & & & \\
0 & 0 & 13l & -3l^2 & 0 & 0 & 0 & 0 & 22l & 4l^2 & & \\
0 & 0 & 0 & 0 & 70 & 0 & 0 & 0 & 0 & 0 & 140 & \\
0 & 0 & 0 & 0 & 0 & \dfrac{70I_{\mathrm p}}{A} & 0 & 0 & 0 & 0 & 0 & \dfrac{140I_{\mathrm p}}{A}
\end{bmatrix}
\qquad (8.21)$$

where $I_{\mathrm p}$ is the polar moment of inertia
    A is the cross-sectional area
    ρ is the density
    l is the element length

The degrees of freedom are in the order:

$$\left\{ w_1, \left(\frac{dw}{dx}\right)_1, v_1, \left(\frac{dv}{dx}\right)_1, u_1, \gamma_1, w_2, \left(\frac{dw}{dx}\right)_2, v_2, \left(\frac{dv}{dx}\right)_2, u_2, \gamma_2 \right\}$$

Rotations

Usually the element matrices are referred to the local frame rather than to the global frame and we need to transform the nodal unknowns. If we use an asterisk to indicate the global frame, we have:

$$\mathbf{U}_e = \mathbf{R}\,\mathbf{U}_e^*$$
$$\delta\mathbf{U}_e = \mathbf{R}\,\delta\mathbf{U}_e^* \tag{8.22}$$

where $\mathbf{R}$ contains the direction cosines. The following element matrix relationship can now be written:

$$\delta\mathbf{U}_e^T\,\mathbf{K}\,\mathbf{U}_e = \delta\mathbf{U}_e^{*,T}(\mathbf{R}^T\mathbf{K}\,\mathbf{R})\mathbf{U}_e^* = \delta\mathbf{U}_e^{*,T}\,\mathbf{K}^*\mathbf{U}_e^*$$

$$\delta\mathbf{U}_e^T\mathbf{M}\ddot{\mathbf{U}}_e = \delta\mathbf{U}_e^{*,T}(\mathbf{R}^T\mathbf{M}\mathbf{R})\mathbf{U}_e^* = \delta\mathbf{U}_e^{*,T}\mathbf{M}^*\,\mathbf{U}_e^* \tag{8.23}$$

$$\delta\mathbf{U}_e^T\mathbf{F} = \delta\mathbf{U}_e^{*,T}\mathbf{R}^T\mathbf{F} = \delta\mathbf{U}_e^{*,T}\,\mathbf{F}^*$$

The governing equations for the *whole* body can be written as:

$$\sum_{n_e}\delta\mathbf{U}_e^T(\mathbf{K}\,\mathbf{U}_e + \mathbf{M}\ddot{\mathbf{U}}_e) = \sum_{n_e}\delta\mathbf{U}_e^T\mathbf{F} \tag{8.24}$$

where n_e = number of elements.

If N denotes the total number of nodes we can define a system nodal unknown vector. (In what follows we assume the $\mathbf{U}_e$ are referred to the global frame, that is we drop the asterisk for simplicity.)

$$\mathbf{U} = \{\mathbf{U}_1, \mathbf{U}_2 \ldots \mathbf{U}_N\} \tag{8.25}$$

Expanding (8.24) by summing the contributions of the elements incident on each node, we have:

$$\delta\mathbf{U}^T\{\mathscr{K}\mathbf{U} + \mathscr{M}\ddot{\mathbf{U}}\} = \delta\mathbf{U}^T\mathscr{F} \tag{8.26}$$

or for arbitrary $\delta\mathbf{U}$:

$$\mathscr{K}\mathbf{U} + \mathscr{M}\ddot{\mathbf{U}} = \mathscr{F} \tag{8.27}$$

The partitioned form of (8.27) is

$$
\begin{bmatrix}
\mathcal{K}_{11} & \mathcal{K}_{12} & \ldots & \mathcal{K}_{1N} \\
\mathcal{K}_{21} & \mathcal{K}_{22} & \ldots & \mathcal{K}_{2N} \\
\ldots & & & \\
\mathcal{K}_{N1} & \mathcal{K}_{N2} & \ldots & \mathcal{K}_{NN}
\end{bmatrix}
\begin{Bmatrix}
\mathbf{U}_1 \\
\mathbf{U}_2 \\
\vdots \\
\mathbf{U}_N
\end{Bmatrix}
+
$$

$$
+
\begin{bmatrix}
\mathcal{M}_{11} & \mathcal{M}_{12} & \ldots & \mathcal{M}_{1N} \\
\mathcal{M}_{21} & \mathcal{M}_{22} & \ldots & \mathcal{M}_{2N} \\
\ldots & & & \\
\mathcal{M}_{N1} & \mathcal{M}_{N2} & \ldots & \mathcal{M}_{NN}
\end{bmatrix}
\begin{bmatrix}
\ddot{\mathbf{U}}_1 \\
\ddot{\mathbf{U}}_2 \\
\vdots \\
\ddot{\mathbf{U}}_N
\end{bmatrix}
=
\begin{Bmatrix}
\mathcal{F}_1 \\
\mathcal{F}_2 \\
\vdots \\
\mathcal{F}_N
\end{Bmatrix}
\tag{8.28}
$$

We assemble $\mathcal{K}$, $\mathcal{M}$, and $\mathcal{F}$ in *partitioned* form, working with successive elements. One first expands the element nodal unknowns vector $\mathbf{U}_e$ in terms of the vector of nodal displacements $\mathbf{U}_{n_i}$, referred to the *global* numbering system:

$$
\mathbf{U}_e = \begin{Bmatrix} \mathbf{U}_{n_1} \\ \mathbf{U}_{n_2} \end{Bmatrix}
\tag{8.29}
$$

where n_1 and n_2 are the global numbers for nodes 1 and 2.

We partition $\mathbf{K}$, $\mathbf{M}$ and $\mathbf{F}$ for the element consistently with the partitioning of $\mathbf{U}_e$.

$$
\delta \mathbf{U}_e^T \mathbf{K} \mathbf{U}_e = \sum_{i=1}^{s} \delta \mathbf{U}_{n_i}^T \left\{ \sum_{j=1}^{s} \mathbf{K}_{ij} \mathbf{U}_{n_j} \right\}
$$

$$
\delta \mathbf{U}_e^T \mathbf{M} \mathbf{U}_e = \sum_{i=1}^{s} \delta \mathbf{U}_{n_i}^T \left\{ \sum_{j=1}^{s} \mathbf{M}_{ij} \mathbf{U}_{n_j} \right\}
\tag{8.30}
$$

$$
\delta \mathbf{U}_e^T \mathbf{F} = \sum_{i=1}^{s} \mathbf{U}_{n_i}^T \mathbf{F}_i
$$

For the beam $s = 2$, but for other elements it is equal to the number of nodes in the element. The contribution for an element is listed below.

In $\mathcal{F}$: $\mathbf{F}_i$ in row i $i = 1, 2 \ldots s$

In $\mathcal{K}$: $\mathbf{K}_{ij}$ in row i, column j $i, j = 1, 2 \ldots s$ (8.31)

In $\mathcal{M}$: $\mathbf{M}_{ij}$ in row i, column j $i, j = 1, 2 \ldots s$

These operations are carried out for all the elements. Since $\mathbf{M}$ and $\mathbf{K}$ are symmetrical for our applications, the $\mathcal{M}$ and $\mathcal{K}$ matrices of (8.27) are going to be symmetrical, and only the coefficients on and above the diagonal need to be stored.

The displacement boundary conditions have to be imposed in system (8.27) before solving. In dynamic problems this is usually done by suppressing the equations corresponding to the known displacements.

8.3 PLATE ELEMENTS

Let us now describe how two-dimensional plane stress elements can be obtained. We start with the two governing equations:

$$\frac{\partial \sigma_x}{\partial x} + \frac{\partial \tau}{\partial y} + \rho \ddot{u} = 0$$

$$\frac{\partial \sigma_y}{\partial y} + \frac{\partial \tau}{\partial x} + \rho \ddot{v} = 0$$

$$(8.32)$$

The boundary conditions are:

$$\begin{aligned} p_x = \sigma_x l + \tau m = \bar{p}_x \\ p_y = \sigma_y m + \tau l = \bar{p}_y \end{aligned} \quad \text{on } S_2 \quad (8.33)$$

l and m are the direction cosines of the normal to the boundary with respect to the x and y axes. The corresponding weighted residual statement is:

$$\int \left\{ \left(\frac{\partial \sigma_x}{\partial x} + \frac{\partial \tau}{\partial y} + \rho \ddot{u} \right) \delta u + \left(\frac{\partial \sigma_y}{\partial y} + \frac{\partial \tau}{\partial x} + \rho \ddot{v} \right) \delta v \right\} t \, dA$$

$$= \int_{S_2} t \left\{ (p_x - \bar{p}_x) \delta u + (p_y - \bar{p}_y) \delta v \right\} ds \quad (8.34)$$

where t is the thickness of the plate. Integrating by parts we obtain:

$$\int t \{ \sigma_x \delta \varepsilon_x + \sigma_y \delta \varepsilon_y + \tau \delta \gamma + \rho \ddot{u} \delta u + \rho \ddot{v} \delta v \} \, dA = \int t (\bar{p}_x \delta u + \bar{p}_y \delta v) \, ds$$

$$(8.35)$$

Or in matrix form:

$$\int t (\delta \boldsymbol{\epsilon}^T \boldsymbol{\sigma} + \delta \mathbf{u}^T \ddot{\mathbf{u}}) \, dA = \int \delta \mathbf{u}^T \mathbf{p} t \, ds \quad (8.36)$$

Note that the u and v functions are assumed to satisfy the displacement boundary conditions on S_1 ($u = \bar{u}$, and $v = \bar{v}$). The matrices in (8.36) are:

$$\boldsymbol{\epsilon} = \begin{bmatrix} \varepsilon_x \\ \varepsilon_y \\ \gamma_{xy} \end{bmatrix} = \begin{bmatrix} \partial u/\partial x \\ \partial v/\partial y \\ \dfrac{\partial u}{\partial y} + \dfrac{\partial v}{\partial x} \end{bmatrix} \tag{8.37}$$

The corresponding stresses $\boldsymbol{\sigma}$ for the isotropic case are given by:

$$\boldsymbol{\sigma} = \mathbf{D}\boldsymbol{\epsilon}$$

where $$\mathbf{D} = \frac{E}{1-v^2} \begin{bmatrix} 1 & v & 0 \\ v & 1 & 0 \\ 0 & 0 & \dfrac{1-v}{2} \end{bmatrix} \tag{8.38}$$

where E is Young's modulus and v is Poisson's ratio.

The displacements throughout the element can be written as:

$$\mathbf{u} = \begin{Bmatrix} u \\ v \end{Bmatrix} \tag{8.39}$$

The boundary forces are:

$$\mathbf{p} = t \begin{Bmatrix} p_x \\ p_y \end{Bmatrix} \tag{8.40}$$

Since the strains consist of only first-order derivatives, to ensure compatibility between elements we have to ensure continuous u and v displacements between elements. The expression should also be able to represent constant strain states, and in particular zero strains (i.e. the rigid body case).

Triangular element

For a simple triangular element such as the one shown in Figure 8.5 we have:

$$\begin{aligned} u &= \alpha_1 + \alpha_2 x + \alpha_3 y \\ v &= \alpha_4 + \alpha_5 x + \alpha_6 y \end{aligned} \tag{8.41}$$

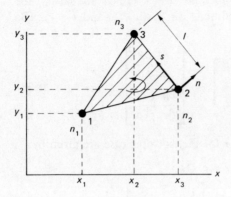

Figure 8.5   Triangular element

These can be written as:

$$\mathbf{u} = \begin{Bmatrix} u \\ v \end{Bmatrix} = \begin{bmatrix} 1 & x & y & 0 & 0 & 0 \\ 0 & 0 & 0 & 1 & x & y \end{bmatrix} \begin{Bmatrix} \alpha_1 \\ \alpha_2 \\ \vdots \\ \alpha_6 \end{Bmatrix}$$

$$= \mathbf{A\alpha} \tag{8.42}$$

Specifically this expression for the three nodes shown in Figure 8.5 becomes:

$$\begin{Bmatrix} u_1 \\ v_1 \\ u_2 \\ v_2 \\ u_3 \\ v_3 \end{Bmatrix} = \begin{bmatrix} 1 & x_1 & y_1 & 0 & 0 & 0 \\ 0 & 0 & 0 & 1 & x_1 & y_1 \\ 1 & x_2 & y_2 & 0 & 0 & 0 \\ 0 & 0 & 0 & 1 & x_2 & y_2 \\ 1 & x_3 & y_3 & 0 & 0 & 0 \\ 0 & 0 & 0 & 1 & x_3 & y_3 \end{bmatrix} \begin{Bmatrix} \alpha_1 \\ \alpha_2 \\ \alpha_3 \\ \alpha_4 \\ \alpha_5 \\ \alpha_6 \end{Bmatrix} \tag{8.43}$$

i.e.
$$\mathbf{U}_e = \mathbf{C\alpha}$$

Hence we can find:

$$\mathbf{\alpha} = \mathbf{C}^{-1}\mathbf{U}_e \tag{8.44}$$

Substituting (8.42), we find:

$$\mathbf{u} = (\mathbf{AC}^{-1})\mathbf{U}_e = \mathbf{GU}_e \tag{8.45}$$

This can be written in full as:

$$u = g_1 u_1 + g_2 u_2 + g_3 u_3$$
$$v = g_1 v_1 + g_2 v_2 + g_3 v_3$$

(8.46)

where

$$g_1 = \frac{1}{2A}(2A_1^0 + b_1 x + a_1 y) \qquad g_2 = \frac{1}{2A}(2A_2^0 + b_2 x + a_2 y)$$

$$g_3 = \frac{1}{2A}(2A_3^0 + b_3 x + a_3 y)$$

with

$$a_i = x_k - x_j \qquad b_i = y_i - y_k \qquad 2A_i^0 = x_j y_k - x_k y_j$$
$$i = 1, 2, 3 \quad j = 2, 3, 1 \quad k = 3, 1, 2 \quad 2A = b_1 a_2 - a_1 b_2$$

Once these expressions are found we can calculate the strains, i.e.

$$\boldsymbol{\epsilon} = \left\{ \begin{array}{c} \varepsilon_x \\ \varepsilon_y \\ \gamma \end{array} \right\} = \begin{bmatrix} \dfrac{\partial g_1}{\partial x} & 0 & \dfrac{\partial g_2}{\partial x} & 0 & \dfrac{\partial g_3}{\partial x} & 0 \\[2mm] 0 & \dfrac{\partial g_1}{\partial y} & 0 & \dfrac{\partial g_2}{\partial y} & 0 & \dfrac{\partial g_3}{\partial y} \\[2mm] \dfrac{\partial g_1}{\partial y} & \dfrac{\partial g_1}{\partial x} & \dfrac{\partial g_2}{\partial y} & \dfrac{\partial g_2}{\partial x} & \dfrac{\partial g_3}{\partial y} & \dfrac{\partial g_3}{\partial x} \end{bmatrix} \mathbf{U}_e \quad (8.47)$$

or

$$\boldsymbol{\epsilon} = \mathbf{B} \mathbf{U}_e$$

We can now calculate the element matrices, which are:

$$\mathbf{K} = \int t EI \mathbf{B}^T \mathbf{B} \, dA$$

$$\mathbf{M} = \int t \mathbf{G}^T \rho \mathbf{G} \, dA$$

(8.48)

$$\mathbf{F} = \int \mathbf{G}^T \mathbf{p} \, ds$$

Rectangular element

For a rectangular element such as the one shown in Figure 8.6 we can take:

$$u = \alpha_1 + \alpha_2 x + \alpha_3 y + \alpha_4 xy$$
$$v = \alpha_5 + \alpha_6 x + \alpha_7 y + \alpha_8 xy \tag{8.49}$$

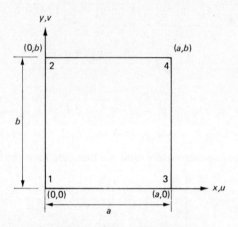

Figure 8.6   Rectangular plate element

After specialising these expressions, we find that:

$$u = (1 - \bar{x})(1 - \bar{y})u_1 + (1 - \bar{x})\bar{y}u_2 + \bar{x}(1 - \bar{y})u_3 + \bar{x}\bar{y}u_4$$
$$v = (1 - \bar{x})(1 - \bar{y})v_1 + (1 - \bar{x})\bar{y}v_2 + \bar{x}(1 - \bar{y})v_3 + \bar{x}\bar{y}v_4 \tag{8.50}$$

where $\bar{x} = x/a$ and $\bar{y} = y/b$. For the element shown in Figure 8.6 we can write these equations as:

$$u = H_1(\bar{x})H_1(\bar{y})u_{11} + H_1(\bar{x})H_2(\bar{y})u_{12} + H_2(\bar{x})H_1(\bar{y})u_{21} +$$
$$+ H_2(\bar{x})H_2(\bar{y})u_{22} \tag{8.51}$$

(and similarly for v), where $H_1(\bar{x}) = 1 - \bar{x}$ and $H_2(\bar{x}) = \bar{x}$. H_1 and H_2 are simple examples of interpolation functions of the type used in plate bending. They have the following properties:

$$H_1(0) = H_2(1) = 1$$
$$H_1(1) = H_2(0) = 0$$

(8.52)

A typical discretisation for a plate is shown in Figure 8.7. We take nodes at the corner of the elements on the middle surface of the plate. The element connectivity table is shown in the figure. Note that the nodes have to be listed in the same direction, but it is irrelevant which is the starting node.

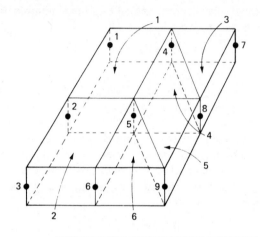

Element	Node			
	n_1	n_2	n_3	n_4
1	1	2	5	4
2	2	3	6	5
3	4	8	7	–
4	4	5	8	–
5	5	9	8	–
6	5	6	9	–

Figure 8.7   Plate divided into six elements

The assemblage of all the elements is carried out as shown previously, with the difference that s is now equal to 3 for the triangular element or 4 for the rectangular one.

8.4 SYSTEM RESPONSE

The equations of motion for the multi-degree-of-freedom system can be expressed as:

$$\mathcal{K}\mathbf{U} + \mathcal{M}\ddot{\mathbf{U}} = \mathcal{F} \tag{8.53}$$

or, if viscous damping is also included:

$$\mathcal{K}\mathbf{U} + \mathcal{C}\dot{\mathbf{U}} + \mathcal{M}\ddot{\mathbf{U}} = \mathcal{F} \tag{8.54}$$

Structural damping is usually represented by a linear combination of the stiffness and mass matrices, i.e.

$$\mathcal{C} = \alpha\mathcal{K} + \beta\mathcal{M} \tag{8.55}$$

where α and β are chosen in such a way as to approximate as closely as possible to the damping of the actual system. Information about damping is usually presented as a percentage of critical damping per mode of vibration.

In general the damping matrix $\mathcal{C}$ is formed by the structural and the hydrodynamic damping. The mass matrix also consists of the contribution from the material and the hydrodynamic mass.

For forced vibrations one can use complex algebra by taking the Fourier transform of (8.54), i.e.

$$\{\mathcal{K} + i\omega\mathcal{C} - \omega^2\mathcal{M}\}\overline{\mathbf{U}} = \overline{\mathcal{F}} \tag{8.56}$$

or $$\overline{\mathbf{Z}}(\omega)\overline{\mathbf{U}} = \overline{\mathcal{F}}$$

The inverse is:

$$\overline{\mathbf{U}} = (Z^{-1})\overline{\mathcal{F}} = \overline{\mathbf{H}}\overline{\mathcal{F}} \tag{8.57}$$

Similarly, for the conjugate of the Fourier transform:

$$\{\mathcal{K} - i\omega\mathcal{C} - \omega^2\mathcal{M}\}\hat{\mathbf{U}} = \hat{\mathcal{F}} \tag{8.58}$$

or $$\hat{\mathbf{Z}}(\omega)\hat{\mathbf{U}} = \hat{\mathcal{F}}$$

Its inverse is:

$$\hat{\mathbf{U}} = (\hat{Z}^{-1})\hat{\mathcal{F}} = \hat{\mathbf{H}}\hat{\mathcal{F}} \tag{8.59}$$

If the force functions are functions of only one random variable, such as the wave amplitude, we have:

$$\mathcal{F} = \mathbf{A}\,\eta(t) \tag{8.60}$$

Hence the transforms give:

$$\hat{\mathscr{F}} = \mathbf{A}\,\hat{\eta} \qquad \overline{\mathscr{F}} = \mathbf{A}^*\overline{\eta} \tag{8.61}$$

where $\mathbf{A}^*$ is a complex conjugate of $\mathbf{A}$. For a point j the displacement can be written as:

$$\overline{U}_j = \overline{\mathbf{H}}_j^T \overline{\mathscr{F}} \qquad \hat{U}_j = \hat{\mathbf{H}}_j^T \hat{\mathscr{F}} \tag{8.62}$$

where $\mathbf{H}_j^T$ are the jth row of the $\mathbf{H}$ matrices.

One can now write the following relationship between the spectral densities of the random forces and the response:

$$S_{u_i u_j} = \overline{\mathbf{H}}_i^T \mathbf{S}_{FF}\,\hat{\mathbf{H}}_j \tag{8.63}$$

The forces spectral density matrix is related to the wave height using (8.61), i.e.

$$\mathbf{S}_{FF} = \mathbf{A}^T S_{\eta\eta}\mathbf{A}^* = |\mathbf{A}|^2 S_{\eta\eta} \tag{8.64}$$

Hence (8.63) can now be written as:

$$S_{u_i u_j} = (\overline{\mathbf{H}}_i^T |\mathbf{A}|^2 \hat{\mathbf{H}}_j) S_{\eta\eta} = |H_{ij}|^2 S_{\eta\eta} \tag{8.65}$$

When the number of degrees of freedom of the system is high, modal superposition is generally used in preference to the forced vibrations response employed above. For many structures the response is mainly in the first few modes and it is not necessary to take too many modes into consideration. It is then more efficient to orthogonalise the system of equations (8.54) before using spectral density functions.

Starting with (8.53) one can first find the eigenvalues and eigenvectors of the system by solving the homogeneous system:

$$\mathscr{M}\ddot{\mathbf{U}} + \mathscr{K}\mathbf{U} = \mathbf{0} \tag{8.66}$$

which gives the characteristic equation:

$$(\mathscr{K} - \omega_i^2 \mathscr{M})\mathbf{Z}_i = \mathbf{0} \tag{8.67}$$

ω_i and $\mathbf{Z}_i$ are the different eigenvalues and the corresponding eigenvectors. The response of the structure can be written:

$$\mathbf{U} = \mathbf{Z}\mathbf{q} = \sum_{i=1}^{s} q_i \mathbf{Z}_i \tag{8.68}$$

where $\mathbf{Z}$ is the modal matrix and q_i the generalised coordinates. One only considers the first few s generalised coordinates and modes.

We can substitute (8.68) into (8.54) and multiply it by $\mathbf{Z}_i$. The matrices $\mathscr{K}$ and $\mathscr{M}$ are orthogonal with respect to these modes, but $\mathscr{C}$ is generally non-orthogonal owing to the hydrodynamic drag contribution. If we assume it can be orthogonalised, we have that:

$$(\mathbf{Z}^T\mathscr{M}\mathbf{Z})\ddot{\mathbf{q}} + (\mathbf{Z}^T\mathscr{C}\mathbf{Z})\dot{\mathbf{q}} + (\mathbf{Z}^T\mathscr{K}\mathbf{Z})\mathbf{q} = (\mathbf{Z}^T\mathscr{F}) \qquad (8.69)$$

becomes:

$$\mathscr{M}^*\ddot{\mathbf{q}} + \mathscr{C}^*\dot{\mathbf{q}} + \mathscr{K}^*\mathbf{q} = \mathscr{F}^* \qquad (8.70)$$

where the $\mathscr{M}^*$, $\mathscr{C}^*$ and $\mathscr{K}^*$ matrices are now diagonal. This system of equations can also be written as s uncoupled second-order differential equations, i.e.

$$\mathscr{M}_i^*\,\ddot{q}_i + \mathscr{C}_i^*\,\dot{q}_i + \mathscr{K}_i^*\,q_i = \mathscr{F}_i^* \qquad (8.71)$$

where $i = 1, s$.

The matrix $\mathscr{C}_i^*$ can be diagonalised by minimising the error vector $\mathbf{E}$:

$$\mathbf{E} = \mathbf{C}_0\dot{\mathbf{q}} - \mathbf{C}^*\dot{\mathbf{q}} \qquad (8.72)$$

where $\mathbf{C}_0$ is the actual (uncoupled) damping matrix obtained after multiplying by $\mathbf{Z}$, and $\mathbf{C}^*$ is the diagonal matrix that we want to obtain with the least possible error.

$$\mathbf{C}_0 = \mathbf{Z}^T\mathscr{C}\mathbf{Z} \qquad (8.73)$$

The mean error can be minimised by using:

$$\left\langle \frac{\partial E^2}{\partial \mathscr{C}_i^*} \right\rangle = \left\langle \left(\sum_{j=1}^{s} C_{0_{ji}}\dot{q}_j - C_i^*\dot{q}_i \right)\dot{q}_i \right\rangle = 0$$

$$\therefore \qquad \mathscr{C}_i^* = C_{0_{ii}} + \sum_{\substack{j=1 \\ j \neq i}}^{s} \frac{C_{0_{ij}} \langle \dot{q}_i\dot{q}_j \rangle}{\langle \dot{q}_j^2 \rangle} \qquad (8.74)$$

As a first approximation the supposition that $\mathscr{C}_i^* = C_{0_{ii}}$ appears to be reasonable, especially taking into account the fact that the hydrodynamic damping contribution is small by comparison with the structural damping part. The designer can, however, check the validity of this hypothesis using equation (8.74) once the $\dot{q}_i$ values have been obtained.

In order to consider the probabilistic response we can consider the

Fourier transform of equation (8.71), i.e.

$$(-\omega^2 \mathcal{M}_i^* + i\omega \mathcal{C}_i^* + \mathcal{K}_i^*)\bar{q}_i = \bar{\mathscr{F}}_i^*$$

or
$$\bar{Z}_i^*(\omega)\bar{q}_i = \bar{\mathscr{F}}_i^* \tag{8.75}$$

$$\therefore \qquad \bar{q}_i = \bar{H}_i^*(\omega)\bar{\mathscr{F}}_i^*$$

where $\bar{H}_i^*(\omega) = \{\bar{Z}_i^*\}^{-1}$. The complex conjugate Fourier transform gives:

$$(-\omega^2 \mathcal{M}_j^* - i\omega \mathcal{C}_j^* + \mathcal{K}_j^*)\hat{q}_j = \hat{\mathscr{F}}_j^*$$

$$\hat{Z}_j^*(\omega)\hat{q}_j = \hat{\mathscr{F}}_j^* \tag{8.76}$$

$$\therefore \qquad \hat{q}_j = \hat{H}_j^*(\omega).\hat{\mathscr{F}}_j^*$$

where $\hat{H}_j^*(\omega) = \{\hat{Z}_j^*\}^{-1}$. Note that these response functions are the same as those for the one-degree-of-freedom system of Chapter 7.

Multiplying equations (8.75) and (8.76) together and dividing by T (when $T \to \infty$), we obtain:

$$S_{qq}(i,j) = H_i^* S_{FF}^*(i,j)\hat{H}_j^* \tag{8.77}$$

The spectral density matrix $\mathbf{S}_{FF}^*$ of the Fourier-transformed forces is related to the spectral density of the original forces $\mathbf{S}_{FF}$ by the transformation:

$$\mathbf{S}_{FF}^* = \mathbf{Z}^T \mathbf{S}_{FF} \mathbf{Z} \tag{8.78}$$

while the spectral density matrix of the original displacements, $\mathbf{S}_{UU}$, is related to the spectral density of the generalised coordinates by:

$$\mathbf{S}_{UU} = \mathbf{Z} \mathbf{S}_{qq} \mathbf{Z}^T \tag{8.79}$$

Note that we can write:

$$\mathbf{S}_{UU} = \mathbf{Z}(\bar{\mathbf{H}}^T \mathbf{S}_{FF}^* \hat{\mathbf{H}})\mathbf{Z}^T \tag{8.80}$$

Each element of S_{FF}^* can also be written as:

$$S_{FF}^*(i,j) = \frac{1}{2\pi} \int_0^\infty R_{FF}^*(i,j)\exp(-i\omega t)\,\mathrm{d}t \tag{8.81}$$

and the cross-correlation terms are:

$$R_{FF}^*(i,j) = E\left[F_i^*(t)F_j^*(t+\tau)\right] \tag{8.82}$$

If the F^* forces depend on only one variable (such as wave amplitude in offshore structures), we can express all elements of S_{FF} as functions of ω, multiplied by the wave spectral density $S_{\eta\eta}(\omega)$, i.e.

$$\mathbf{S}_{FF}^* = \mathbf{A}S_{\eta\eta}(\omega)\mathbf{A}^T = |\mathbf{A}|^2 S_{\eta\eta} \qquad (8.83)$$

The above equations give the complete solution for the random vibrations of the system, but a common assumption is to neglect the off-diagonal terms in the $\mathbf{S}_{FF}^*$ matrix. The justification for this is that all is well, provided that the natural frequencies are well separated. Unfortunately this assumption may lead to non-conservative results for complex structures, and designers are recommended to use the full analysis.

Example 8.2: force spectral density on a stationary member

From the linearised Morison's equation—equation (4.19)—we have, for a stationary cylinder, the force per unit length at depth z in the direction of wave advance:

$$F(z, t) = C_I \dot{v}_x + C_D \sqrt{(8/\pi)}\sigma_v v_x \qquad (a)$$

The variables $\dot{v}_x$ and v_x are evaluated on the axis of the cylinder, here taken as $x = 0, y = 0$, and for a linear wave of amplitude a_0 in water of depth h we have (see equation (3.50)):

$$\Phi(x, z, t) = +\frac{ig}{\omega} a_0 \exp[-i(\kappa x - \omega t)] \frac{\cosh[\kappa(z + d)]}{\cosh \kappa d} \qquad (b)$$

where $\omega^2 = g\kappa \tanh \kappa d$.

We take the real part of equation (8.83) for the actual physical variable Φ, and we have taken the x axis in the direction of wave advance and hence in the direction of the wave force F.

$$\therefore \quad v_x = \frac{\partial \Phi}{\partial x} = \frac{g\kappa}{\omega} a_0 \exp[-i(\kappa x - \omega t)] \frac{\cosh[\kappa(z + d)]}{\cosh \kappa d} \qquad (c)$$

and

$$\dot{v}_x = \frac{\partial v_x}{\partial t} = ig\kappa a_0 \exp[-i(\kappa x - \omega t)] \frac{\cosh[\kappa(z + d)]}{\cosh \kappa d} \qquad (d)$$

Hence, for this particular wave of frequency ω and corresponding amplitude $a_0(\omega)$, we may obtain the force on the cylinder corresponding to a wave of this frequency per unit length at depth z as:

$$F(z, t; \omega) = \left[g\kappa \, \frac{\cosh\left[\kappa(z+d)\right]}{\cosh \kappa d} \left\{ C_D \sqrt{\left(\frac{8}{\pi}\right)} \frac{\sigma_v}{\omega} + \right. \right.$$

$$\left. \left. + iC_I \right\} \right] \exp\left(i\omega t\right) a_0(\omega) \tag{e}$$

Consider the surface elevation at the cylinder axis $\eta(t)$ to be a random variable; since it is bounded (and continuous) with time, we may obtain the Fourier transform of $\eta(t)$, defined by:

$$\tilde{\eta}(\omega) = \frac{1}{2\pi} \int_{-\infty}^{\infty} \eta(t) \exp\left(-i\omega t\right) dt \tag{f}$$

which implies, by the orthogonality of $\exp\left(+i\omega t\right)$, that:

$$\eta(t) = \int_{-\infty}^{\infty} \tilde{\eta}(\omega) \exp\left(i\omega t\right) d\omega \tag{g}$$

The right-hand side may be thought of as a sum of Fourier components $\tilde{\eta}(\omega) \exp\left(+i\omega t\right)$ that are harmonic waves of amplitude $\tilde{\eta}(\omega)$.

Returning to equation (d) we now see that we can identify $a_0(\omega)$ with this Fourier transform, and in the same way the force $F(z, t; \omega)$ corresponds to the Fourier component of the random force $F(z, t)$ resulting from an incident random wave field $\eta(t)$. Hence we rewrite $F(z, t; \omega)$ as $\tilde{F}(z, \omega) \exp\left(+i\omega t\right)$, so we have:

$$\tilde{F}(z, \omega) \exp\left(i\omega t\right) = \left[g\kappa \, \frac{\cosh\left[\kappa(z+d)\right]}{\cosh \kappa d} \left\{ C_D \sqrt{\left(\frac{8}{\pi}\right)} \frac{\sigma_v}{\omega} + \right. \right.$$

$$\left. \left. + iC_I \right\} \right] \tilde{\eta}(\omega) \exp\left(i\omega t\right) \tag{h}$$

If we now consider the complex conjugate of equation (a), we have:

$$\Phi(x, z, t) = -\frac{ig}{\omega} a_0 \exp\left[i(\kappa x - \omega t)\right] \frac{\cosh\left[\kappa(z+d)\right]}{\cosh \kappa d} = \Phi^*(x, z, t) \tag{i}$$

This complex conjugation does not change the physical situation represented by Φ, as the real part of the right-hand sides of equations

(a) and (h) are equal. Carrying through the analysis for this Φ, we then obtain for the force:

$$
F(z, t; \omega) = \left[g\kappa \frac{\cosh[\kappa(z+d)]}{\cosh \kappa d} \left\{ C_D \sqrt{\left(\frac{8}{\pi}\right)} \frac{\sigma_v}{\omega} - \right.\right.
$$
$$
\left.\left. - iC_I \right\} \right] \exp(-i\omega t)\, a_o(\omega) \tag{j}
$$

In the same way as before we may identify $a_o(\omega) \exp(-i\omega t)$ with the Fourier component $\bar{\eta}(\omega)$ of $\eta(t)$, this time defined by:

$$
\bar{\eta}(\omega) = \frac{1}{2\pi} \int_{-\infty}^{\infty} \eta(t) \exp(i\omega t)\, dt \tag{k}
$$

which gives:

$$
\eta(t) = \int_{-\infty}^{\infty} \bar{\eta}(\omega) \exp(-i\omega t)\, d\omega \tag{l}
$$

Note that $\bar{\eta}(\omega) = \hat{\eta}(\omega)^*$, as $\eta(t)$ is real.

For the same reasons as before we rewrite $F(z, t; \omega)$ as $\bar{F}(z, \omega)$ $\exp(-i\omega t)$, the Fourier component of the random force $F(z, t)$, using the same transform convention as for η. Hence we may write (j) as:

$$
F(z, \omega) \exp(-i\omega t) = \left[g\kappa \frac{\cosh[\kappa(z+d)]}{\cosh \kappa d} \left\{ C_D \sqrt{\left(\frac{8}{\pi}\right)} \frac{\sigma_v}{\omega} - \right.\right.
$$
$$
\left.\left. - iC_I \right\} \right] \bar{\eta}(\omega) \exp(-i\omega t) \tag{m}
$$

Note that $\bar{F}(x, \omega) = \tilde{F}(z, \omega)^*$.

Comparing equations (g) and (l), and using the analysis of Chapter 2, we can consider the expressions in square brackets in both equations to be the 'transfer functions' that relate the wave heights and wave forces. So we may write simply:

$$
\tilde{F}(z, \omega) = T(\omega, z)\, \hat{\eta}(\omega)
$$
$$
\bar{F}(z, \omega) = T(\omega, z)^*\, \bar{\eta}(\omega) \tag{n}
$$

where $T(\omega, z) = g\kappa \dfrac{\cosh[\kappa(z+d)]}{\cosh \kappa d} \left\{ C_D \sqrt{\left(\dfrac{8}{\pi}\right)} \dfrac{\sigma_v}{\omega} + iC_I \right\}$ (p)

and from equation (2.53) the power spectral density of the force per unit length at depth z is given by:

$$S_{FF}(\omega, z) = \lim_{T \to \infty} \frac{1}{T} \tilde{F}(z, \omega)\, \overline{F}(z, \omega)$$

$$= \lim_{T \to \infty} \frac{1}{T} |\tilde{F}(z, \omega)|^2 \tag{q}$$

and
$$S_{\eta\eta}(\omega) = \lim_{T \to \infty} \frac{1}{T} |\tilde{\eta}(\omega)|^2 \tag{r}$$

Therefore equations (n) give:

$$S_{FF}(\omega, z) = |T(\omega, z)|^2\, S_{\eta\eta}(\omega) \tag{s}$$

or
$$S_{FF}(\omega, z) = g\kappa\, \frac{\cosh[\kappa(z+d)]}{\cosh \kappa d} \times$$

$$\times \left[C_D^2\, \frac{8}{\pi}\, \frac{\sigma_v^2}{\omega^2} + C_I^2 \right] S_{\eta\eta}(\omega)$$

Bibliography

Brebbia, C. A., *et al.*, *Vibrations of engineering structures*, Southampton University Press (1974)

Brebbia, C. A., and Connor, J. J., *Fundamentals of finite element techniques for structural engineers*, Butterworths (1973)

Connor, J. J., and Brebbia, C. A., *Finite element techniques for fluid flow*, Newnes–Butterworths (1976)

Malhotra, A. K., and Penzien, J., 'Non-deterministic analysis of offshore structures', *Proc. ASCE (Mech. Div.)*, **96** (1970)

9 Response of Offshore Structures

9.1 INTRODUCTION

In this chapter we study the response of fixed offshore structures of the type used in water depths in the order of 100 to 200 metres, for which the dynamic amplification is significant. These structures can be classified into two types, i.e. gravity and lattice structures. The gravity ones are of many different types but basically they consist of a caisson, three, four or more legs, and a deck (see Figure 9.1). The caisson and legs are made with concrete but the deck structure and top of the legs may be built of steel. Lattice structures are usually made of steel (Figure 9.2) and fixed to the sea floor by piles. Gravity structures instead rest on the sea bed without the need of any special foundation.

Offshore structures are subjected to a series of loads. Forces such as gravity loads, buoyancy, forces induced by the average wind velocity, temperature stresses, certain operational loads and current forces can be considered as static. The most important dynamic forces are those due to the waves. Sometimes even wave forces are considered as static, but for structures with natural periods shorter than two seconds they have to be considered as dynamic loads. The equivalent static loading approach (or quasi-static analysis) is no longer valid as dynamic amplification becomes important. In addition fatigue effects need to be studied in certain 'hot spots' (mainly in steel structures), and they can only be properly analysed using dynamic theory. Finally,

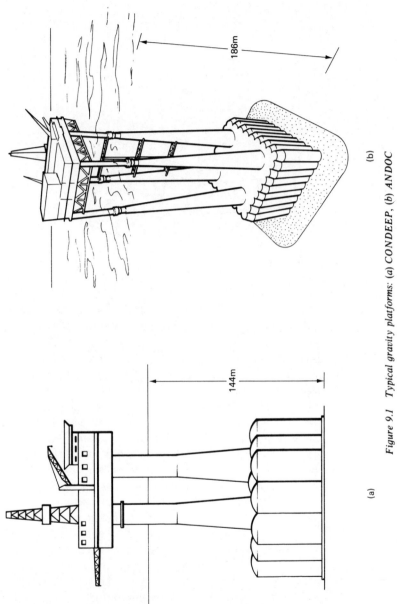

Figure 9.1 Typical gravity platforms: (a) *CONDEEP,* (b) *ANDOC*

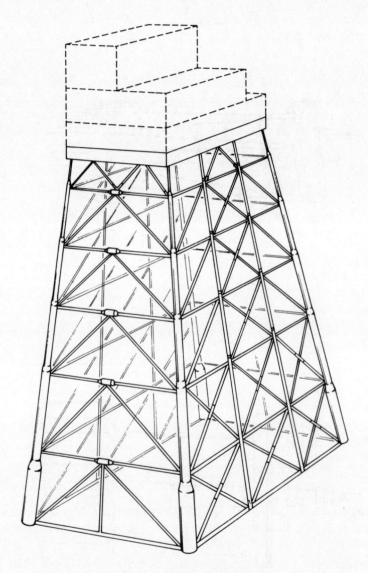

Figure 9.2 Steel lattice structure

foundation–structure interaction effects are generally frequency dependent, as we shall see at the end of this chapter.

If one uses a series of design waves of given height and period, the solutions thus obtained do not represent well the variability of the sea states. This is specially important for fatigue analysis, where a good history prediction of the stress magnitude and number of cycles is needed. It is more accurate in this case to use random vibration analysis, for which a sea spectrum is required.

In what follows we first look at a single column, to provide some comparison between the effect of drag versus inertia and Morison versus diffraction forces. Then the case of a typical gravity-type platform, with and without diffraction, is discussed. Section 9.4 is dedicated to the steel lattice-type platform, and section 9.5 to the fatigue problem.

9.2 THE SINGLE COLUMN

Consider a column structure of the type shown in Figure 9.3, where the column is a concrete cylinder of external diameter D. The water depth is d and the mass of the deck is assumed to be concentrated at a distance l from the sea bed.

The system can be analysed using Morison's equation and diffraction theory. For the first case one can consider the effect of drag, which is neglected when using diffraction theory.

Morison's equation for a cylindrical member produces a force spectrum per unit length given by equation (5.80), i.e.

$$S_{FF}^{m}(\omega, z) = \frac{1}{4}\rho^2 D^2 \pi^2 g^2 \kappa^2 \frac{\cosh^2[\kappa(z+d)]}{\cosh^2 \kappa d} S_{\eta\eta}(\omega) +$$
$$+ \frac{4}{\pi} D^2 \sigma_v^2 \frac{g^2 \kappa^2}{\omega^2} \frac{\cosh^2[\kappa(z+d)]}{\cosh^2 \kappa d} S_{\eta\eta}(\omega) \tag{9.1}$$

where the two terms represent the inertia and drag contributions respectively. When diffraction theory is used we neglect the drag term and find that the inertia contribution becomes (see (5.79)):

$$S_{FF}(\omega, z) = \frac{16\rho^2 g^2}{\kappa^2} \frac{\cosh^2[\kappa(z+d)]}{\cosh^2 \kappa d} \frac{1}{|H_1^{(2)\prime}(\frac{\kappa D}{2})|^2} S_{\eta\eta}(\omega) \tag{9.2}$$

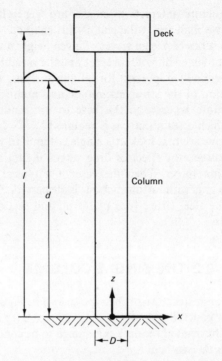

Figure 9.3 Single column

where $H_1^{(2)'}$ is the derivative of the Hankel function.

The ratio between the inertia part of (9.1) and (9.2) is an indication of the relative magnitude of the forces obtained by the two theories. This ratio, γ, was discussed in section 5.3 and plotted in Figure 5.11.

No special difficulty is encountered in using the force spectrum derived from diffraction theory. As the results may be quite different, it is recommended that the proper diffraction spectrum be used for large-diameter members (i.e. those with $D > 6$ metres).

Example 9.1

The following example was obtained from Ebert[1] and represents a single-leg concrete platform with the dimensions shown in Figure 9.4.

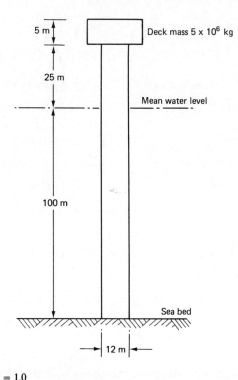

$c_d = 1.0, \quad c_m = 1.0$
Density of concrete $\rho_c = 2.503 \times 10^3 \, \text{kg/m}^3$
Cross-sectional area $A_c = 18.1 \, \text{m}^2$
Mass per unit length $\rho_c A_c = 4.53 \times 10^4 \, \text{kg/m}$
Moment of inertia $I = 380 \, \text{m}^4$
Displaced volume of water per unit length $= 113 \, \text{m}^3/\text{m}$
Structural critical damping ratio $\gamma = 0.05$
Modulus of elasticity $E = 0.16 \times 10^{11} \, \text{N/m}^2$

Figure 9.4   Single-leg structure

The Pierson–Moskowitz spectrum defines the sea state under consideration, i.e.

$$S_{\eta\eta}(\omega) = \frac{A}{\omega^5} \exp\left(-\frac{B}{\omega^4}\right) \qquad \text{(a)}$$

with $A = 2.56$ and $B = 0.0299$.

Results were obtained for the frequencies and modal shapes (see Figure 9.5) corresponding to free vibrations. As a comparison the

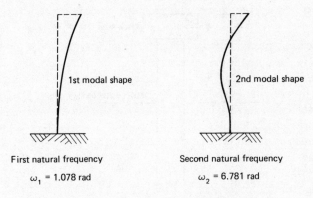

First natural frequency

$\omega_1 = 1.078$ rad

Second natural frequency

$\omega_2 = 6.781$ rad

Figure 9.5   Modal shapes and natural frequencies

Table 9.1   DISPLACEMENT STANDARD DEVIATION AT THE TOP (m).

Including drag in Morison's equation	Neglecting drag	
	Morison's equation	Diffraction
1.430	1.245	0.643

standard deviations of the horizontal displacement at the top of the structure are presented in Table 9.1, for the two cases of including and neglecting drag forces. For the case without drag, two results were obtained, the first using Morison's formula and the second using diffraction theory.

The dimensions and properties of the column are somewhat unrealistic, but were chosen in order to have a large dynamic amplification. Comparison of the results gives some idea of the relative importance of the different terms.

The first point to note is that drag terms (linearised drag is the only effect considered here) are relatively unimportant if one compares the σ obtained using Morison's equation both with and without drag (i.e. 1.43 m to 1.245 m). As expected for a cylinder of large diameter, the response associated with Morison's equation is larger than the one due to diffraction theory if the first natural period of the structure is shorter than the peak period of the spectrum. The large numerical difference (0.643 m against 1.245 m for the standard deviation) for this case is due to the particular dimensions of the system.

9.3 MULTI-LEG GRAVITY STRUCTURES

Diffraction effects are important for large-diameter multi-leg plat-forms. In addition the presence of each leg affects the wave field on the others. The interaction between legs cannot be taken into account using the standard Morison equation, but can be considered as discussed in section 5.4, where the wave field reflected from the other columns is also included.

Interaction effects will appear for column separation of less than one wavelength, and can be especially important at distances of $\frac{1}{3}$, $\frac{2}{3}$ and 1 wavelength, with proportional changes in expected forces varying from $+15$ per cent to -10 per cent approximately. Figures 5.17 and 5.18 show these effects; they are valid for two cylinders, one behind the other with respect to the wave direction. The effect will be less when the pair of cylinders is at an angle to the wave direction.

The forces on the caisson in the horizontal and vertical directions have been calculated in section 5.5. The horizontal force is especially important, and in Example 5.1 it was calculated to be eight times greater than the force on a single cylindrical leg. The corresponding horizontal diffraction coefficient for the base is frequency dependent, and it is dangerous to use a constant over the whole range of frequencies of interest, as is usually done. Notice that this coefficient is given by equation (5.6) as a fourth-power function of frequency. Similar considerations apply to the coefficients for the vertical forces and moments.

Currents produce steady drag forces on the structure and may undermine the foundations through scouring (see Chapter 6).

Example 9.2

The gravity structures can be analysed using three different methods. They are, in order of sophistication:
- equivalent static analysis, sometimes also called 'quasi-static';
- deterministic dynamic analysis;
- random vibrations analysis.

To compare the results that can be obtained in each case, the three methods were applied to study a typical gravity platform, such as the one shown in Figure 9.6. The structure resembles an ANDOC

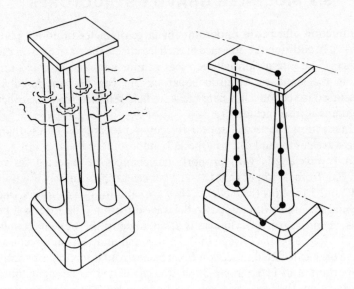

Figure 9.6   General view and discretisation

platform and most of the dimensions are the same as those for the platform shown in Figure 9.1(b). An important distinction is that the deck of the platform and the ends of the columns are assumed to be made of concrete, and not of steel as is the case in the ANDOC structure. This simplifies the problems encountered in the analysis and makes comparisons easier. (For a full discussion of the problems relating the concrete–steel system see Vughts and Kinra[2].)

The four columns of the structure are based on a large 'caisson', and the structure can be idealised as a two-dimensional frame as shown in Figure 9.7. Node 1 is assumed not to move in the horizontal and vertical directions, and the vertical movement of nodes 2 and 3 is neglected. For the purpose of comparison, Morison's formula is assumed throughout, without a special diffraction coefficient for the base.

The columns have an external and internal diameter of 6 and 5 metres respectively at the top. These dimensions are considered to be the same until nodes 12 and 13, and then vary linearly to 15 m (external diameter) and 13 m (internal diameter) at the join between columns and caisson. The distance between column-axes is 40 m.

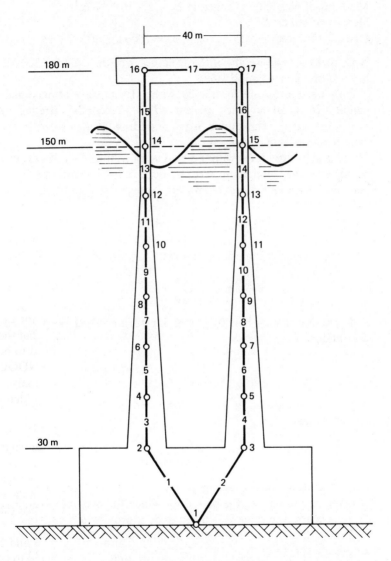

Figure 9.7 Discretisation of platform

The material properties are:

Modulus of elasticity of concrete $E_c = 2 \times 10^{10}$ N/m^2
Density of concrete $\rho_c = 2500$ kg/m^3
Density of water $\rho_w = 1000$ kg/m^3

Note that the total weight of the structure is of the order of 300 000 tonnes, which is a typical value.

Only wave forces are considered, with a design wave of maximum height $H_m = 20$ m and period $T_z = 13$ seconds (frequency $\omega = 0.483$ rad/s). The added mass and drag coefficients for the cylindrical legs are $c_m = c_d = 1.0$.

For gravity structures with these dimensions the effect of drag can be neglected by comparison with the inertia effects. Hence the forces per unit length along the column can be expressed simply by:

$$F(t) = \left(\rho \frac{\pi D^2}{4} + \rho A \right) \dot{v} = 2 \rho \dot{v} \qquad (a)$$

where the water particle accelerations are given by:

$$\dot{v} = a \omega^2 \frac{\cosh \kappa z}{\cosh \kappa d} \sin (\kappa x - \omega t) \qquad (b)$$

If the maximum wave height H_m is acting during a three-hour period, we obtain:

$$H_s = \frac{H_m}{\sqrt{[0.5 \ln (T/T_m)]}} = 10.91 \text{ m} \qquad (c)$$

where T is the duration of the storm and T_m for a narrow-band process is given by:

$$T_m = \frac{T_0}{1.41} \simeq \frac{T_z}{1.41} \qquad (d)$$

T_m is the mean zero crossing period, T_z the most probable zero crossing period and T_0 the period for which the maximum value of $S_{\eta\eta}(\omega)$ occurs. The value T_z is taken to be approximately equal to T_0.

The random sea is represented by the Pierson–Moskowitz wave amplitude spectrum:

$$S_{\eta\eta}(\omega) = \frac{A}{\omega^5} \exp\left(-\frac{B}{\omega^4} \right) \qquad (e)$$

with

$$A = 4B\left(\frac{H_s}{4}\right)^2 = 2.03 \qquad B = \frac{5}{4}\left(\frac{2\pi}{T_0}\right)^{\frac{1}{4}} = 0.068$$

The origin of the x axis is taken to be the left-hand column and the maximum accelerations components for the static case are obtained taking $t = 3.25$ s in formula (b). The dynamic analyses are carried out by determining the first few eigenvalues and eigenvectors of the free vibrating system. The mass of the platform is taken as 18 000 tonnes. The first natural frequency is 1.74 rad/s with a period 3.6 seconds. This value seems to be in agreement with the period observed in concrete gravity structures.

For the forced vibrations analysis the hydrodynamic and structural damping are considered as 1 per cent and 3 per cent of the critical damping, respectively. The random vibration analysis is carried out as indicated in section 9.2.

It is interesting to note that the transfer function curve for horizontal displacement has a series of zero points (Figure 9.8) corresponding to values of λ that are multiples 0.5, 1.5, 2.5, etc. of the distance between columns. (Note that $\omega = 2\pi/T = \sqrt{(2\pi g/\lambda)}$, where

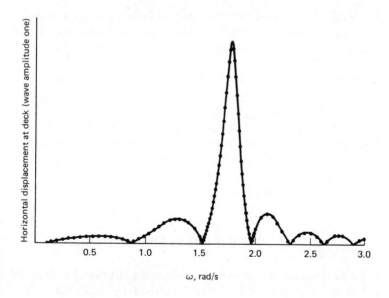

Figure 9.8   Typical displacement response curve for a two-column platform

λ is the wavelength.) For these cases the forces on both columns are equal and opposite and the deck displacement is zero.

The results obtained for the maximum horizontal deck displacements are as follows:

Equivalent static analysis (quasi-static)   0.55 m

Deterministic dynamic analysis   0.75 m

Probabilistic dynamic analysis   1.13 m

For the probabilistic results a value $\lambda = 4$ was taken. A more consistent way of obtaining λ is to find the maximum expected value of the response, i.e.

$$\langle |\bar{U}|_{max} \rangle \simeq \sigma_{\bar{U}\bar{U}} \{ \sqrt{[2 \ln (T/T_m)]} \} \tag{f}$$

where T is the duration of the storm and T_m is now the mean zero crossing period of the displacement response, given by:

$$T_m = 2\pi \left\{ \frac{\int_0^\infty S_{\bar{U}\bar{U}} \, d\omega}{\int_0^\infty \omega^2 \, S_{\bar{U}\bar{U}} \, d\omega} \right\}$$

The results clearly show the importance of dynamic magnification for this type of structure and the fact that a static equivalent analysis (or even a dynamic but deterministic one) can produce non-conservative results.

The above results were obtained using Morison's equation, neglecting the fact that the dimensions of the columns are such that the incident waves are going to be disturbed.

If diffraction theory were taken into account we would obtain a similar effect as for the single-leg structure, with the top displacement deviation decreasing by about 50 per cent for this particular example. Sheltering is not so important and produces a small percentage change (less than 5 per cent).

9.4 LATTICE-TYPE STRUCTURES

If the diameters of the structural members are less than a fifth of the wavelength one can use the classical Morison equation, taking into consideration drag effects. In lattice-type structures this is usually the

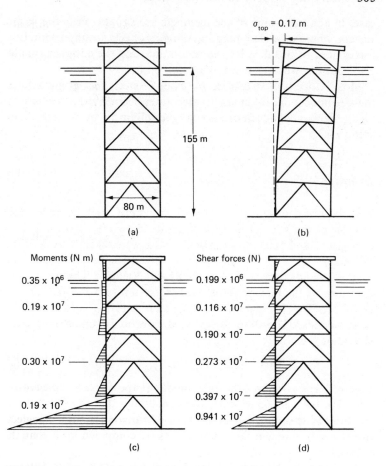

Wave spectrum	Displacement standard deviation at top σ_u, metres		Fundamental natural frequency ω_1, rad/s	Second natural frequency ω_2, rad/s
	Incl. drag	No drag		
$H_s = 18.5\,\text{m}$ $T_m = 11.35\,\text{s}$	0.1695	0.0626	38.054	10.714

Figure 9.9 Steel lattice structure: (a) general configuration, (b) displacement shape, (c) bending moments in legs (including drag), (d) shear forces (including drag)

case. In addition some of the members are situated near the water surface, where wave slamming and slapping effects are important. Lift forces need to be taken into consideration, and the designer should also consider whether vortex shedding occurs.

Most lattice-type structures are made of steel, and fatigue effects that are not significant in gravity structures become a decisive design factor. Because of its importance fatigue will be discussed in detail in section 9.5.

Example 9.3

The following example is taken from Ebert[1] and describes the analysis of the steel lattice-type structure shown in Figure 9.2. The dimensions and properties of the structure were taken directly from the blueprints. A two-dimensional discretisation is shown in Figure 9.9 and consists of 36 elements and 20 nodes. The sea state is described by the Pierson–Moskowitz spectrum with $B = 0.0229$ and $A = 2.56$. (This corresponds to a sea state with $H_m = 30$ m and $T_z = 17$ seconds.) The drag forces need now to be taken into consideration, i.e.

$$F(t) = (C_M + C_A)\dot{v} + C_D\sqrt{(8/\pi)}\sigma_v v \tag{a}$$

Figure 9.9 also shows the first modal shape for a frequency of 0.605 Hz, with a 1.65 second period, assuming that the structure is built-in at the mudline. The standard deviations for the horizontal displacement moments and shear forces are computed for a value of $\gamma_s = 0.05$.

The fundamental period of this type of structure is lower than the period of a gravity-type platform of similar dimensions. This may lead us to believe that the random vibrations analysis is not needed since the dynamic amplification would not be too significant. However, for these structures a probabilistic analysis is important to predict fatigue.

9.5 FATIGUE ANALYSIS

With our present knowledge, the fatigue analysis of offshore structures gives only a rough estimate of system behaviour. This is because of the many approximations involved in the design process. A fatigue analysis needs the following three parts:

● the statistics of the sea, presented in the form of wave height exceedance diagrams or spectral curves;

● the relationship between the sea state and the stresses at the points under consideration (called 'hot spots');

● the determination of the 'stress versus number of cycles' fatigue curves (called S–N curves).

In addition the stress concentration factors may be needed. (This depends on the type of fatigue curve used.) The S–N curves obtained for the hot spot are compared with the S–N curves to failure given by the norms. This is done by using Miner's cumulative damage rule, i.e.

$$\text{Cumulative damage} = (D) = \sum_i \frac{n_i}{N_i} \qquad (9.3)$$

where i = stress range, n_i = number of cycles occurring at stress range i, N_i = number of cycles needed for failure at stress range i. If (D) is accumulated damage occurring during one year, the total life of the structure is:

$$\text{Life (in years)} = \frac{1}{(D)_1} \qquad (9.4)$$

The methods to estimate the fatigue life can be deterministic or probabilistic (Figure 9.10). The deterministic approach is based on knowing the wave height exceedance diagram, usually for one year. Then the relationship between stresses, at the 'hot spot', and wave height is found, and finally the cumulative stress damage curve is computed by plotting the number of cycles against stress range. The number of cycles can be obtained from the wave height exceedance curve. It is generally recommended that the dynamic amplification factor be taken into account. The cumulative stress history curve thus obtained is compared against standard fatigue failure curves. The process is relatively simple and straightforward.

In the probabilistic fatigue analysis one starts by considering the different sea states for a one-year period and the percentage of time during which they act. The number of times that a sea state produces peaks that exceed a particular stress level can be computed from the distribution of stress peaks. The probability curve can then be plotted for each sea state against the stress load. This, and the knowledge of

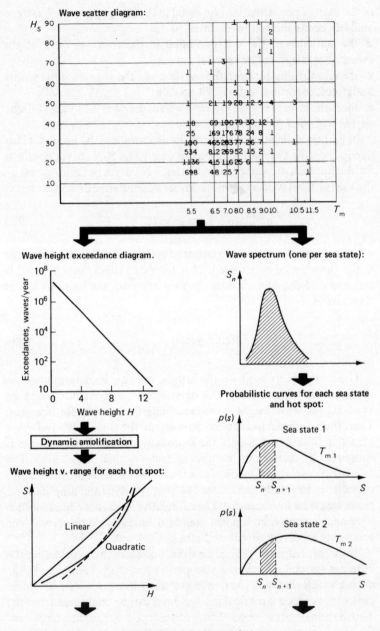

Figure 9.10 Fatigue analysis

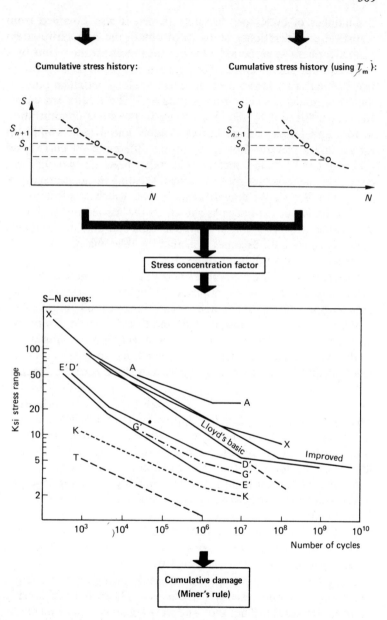

Figure 9.10   continued

the number of cycles per sea state (which is also obtained from statistical considerations), allow us to obtain the cumulative stress history for a one-year period. One can then multiply the results by a stress concentration factor and compare them against the S–N curves using Miner's rule. Hence the fatigue damage for a particular point or 'hot spot' under consideration is obtained. The results are unfortunately influenced by the inaccuracies in the stress concentration factors and fatigue design curves. Vughts and Kinra[2] computed fatigue life at several 'hot spots', using different S–N curves and different concentration factors. The ratio between two different curves for the same hot spot (American Welding Society category X curve and British Welding Institute curve, which is substantially Lloyd's basic but continues as shown by the dashed line in Figure 9.10) varied between 7 and 30. It is concluded that until more reliable high-frequency data becomes available the dilemma of which S–N curve to use will be difficult to resolve.

Stress concentration factors are impossible to determine in a simple form, and although some empirical formulae exist the difference between applying one or the other or carrying out a full finite element analysis can be considerable; Kallaby and Price[3] report differences of up to 100 per cent. This factor has great influence in the results, owing to the exponential character of the S–N curves. In other words the relationship between number of cycles and stress is of the type:

$$N = mS^n \qquad (9.5)$$

where n is the slope of the curve approximately between -2 to -4. Assuming $n = -4$, for instance, a difference of 2 in the stress concentration factors produces a ratio of 16, i.e. fatigue life is 16 times longer for the case with the smaller concentration factors.

Another important point is the directionality of the waves. Using directional spectra yields a substantial reduction in computed forces. Directional spreading is essential for the lower waves of primary interest in fatigue analysis, and Marshall[4] concluded that taking it into account the stress may be influenced by a factor of 2 and the fatigue damage ratio by a factor of 10.

The *deterministic* approach is based on knowing the wave height exceedance diagram, usually for one year. Then the relationship between stresses at the 'hot spot' and wave height is found, and finally a cumulative stress damage curve is computed by plotting number of cycles versus stress range. The number of the cycles can be obtained from the wave height exceedance curve. It is generally recommended that the dynamic amplification factor be taken into account, and sometimes this is obtained by computing the dynamic response of a simplified system instead of the response of the multi-degree-of-freedom structure. The cumulative stress history curve thus obtained is compared with curves such as the AWS or Lloyd's failure curves using Miner's rule.

The process is relatively simple and straightforward. By fitting closed-form expressions for the curves, some authors[5, 6] have deduced formulae for the accumulated fatigue damage. Nolte and Hansford[5] deduced two formulae: one for the fatigue damage resulting from a single sea state (storm) and the other for that occurring during the service life of the structure. The difference is that the maxima for a single sea state can be approximated by a Rayleigh distribution, while a more general Weibull distribution is applied for long-term damage. Several types of deterministic fatigue analysis can be proposed but they are all related to the scheme shown in Figure 9.10, their differences being in the degree and type of approximation used. It is important to keep in mind that each simplification implies another approximation in the results and that simplifications can easily introduce large errors, owing to the exponential character of the curves.

Another shortcoming of the analysis is the difficulty of knowing the zero crossing period for the stress histories, from which the number of stress cycles can be calculated. This period depends both on the structure period and on the period of the exciting waves. The period is computed for the probabilistic fatigue analysis by taking spectral moments, but has to be approximated for the deterministic analysis. The rule is to use the same period as the period of the waves for structures with negligible dynamic amplification. When the dynamic effects cannot be neglected (i.e. for those waves whose frequencies are within a certain interval of the fundamental frequency of the structure) the period is taken as that of the structure.

For a *probabilistic* fatigue analysis one starts by considering the different sea states for a one-year period and the percentage of times during which they act. The number of times a sea state produces peaks that exceed a particular stress level can be computed from the distribution of stress peaks. For each sea state this distribution is fully defined by knowing σ_s (stress deviation at the point or 'hot spot' under consideration). The probability curve can be plotted for each sea state against the stress level S (Figure 9.10).

The number of cycles for the sea state is calculated by first determining the mean stress period:

$$T_{ms} = 2\pi \left\{ \frac{\int_0^\infty S_{ss}\, d\omega}{\int_0^\infty S_{ss}\omega^2\, d\omega} \right\}^{\frac{1}{2}}$$ (9.6)

Hence the number of cycles is:

$$N_s = T/T_{ms}$$ (9.7)

where T is the duration of the sea state under consideration.

Note that the total area under the probability curves in Figure 9.10 is always equal to 1. The shaded area represents the proportion of cycles for a particular stress range; this proportion is then multiplied by the total number of cycles in one year for the given sea state, to give the number of cycles for the particular stress range under consideration. The same is done for all the other cycles, and we obtain a cumulative stress history curve for a one-year period. One can then multiply the cumulative stress history by the stress concentration factor and the total number of years the structure is designed to last (or by twice that number of years as a safety factor).

Finally, Miner's rule is used to estimate the fatigue damage in the particular point or 'hot spot' under consideration.

9.6 SOIL–STRUCTURE INTERACTION

An area of still greater uncertainty than fatigue is that of soil–structure interaction. The main difficulty is the characterisation of the soil, which presents a complex non-linear behaviour. In some cases this behaviour is approximated by an equivalent hysteretic damping (5 to 15 per cent). In addition the loss of energy through the boundaries extending to infinity is taken into account by radiation damping, and this is usually estimated using half-space theories. The half-space results[7] used have the disadvantage of being valid only for homogeneous elastic soils, for which no material damping occurs. Recent studies allow the extension of the theory to account for damping, but the results are not in tabulated form. In addition it has been found that a layered medium behaves in a different way from a homogeneous one, resulting in a stronger frequency dependency for the radiation damping[8].

All these difficulties occurring with half-space theories have led some researchers to use finite-element representations for the soil. Although finite elements allow for a better variation of the soil properties, they tend to represent radiation damping inaccurately and have problems connected with mesh resolution. Radiation damping is very significant for soils, but finite element discretisations are unable to represent it properly owing to the approximations made in the boundary conditions. Special elements for the boundaries have been developed but there is still controversy regarding their performance. Care should be taken to discretise the region with a mesh fine enough not to produce artificial filtering of frequencies. The mesh size is a function of the velocity of wave propagation or celerity. The wavelength can be defined as:

$$\lambda = TV_s \tag{9.8}$$

where T is the period of the exciting frequency for steady-state oscillations and V_s is the velocity of propagation of the wave. Typical wavelengths are shown in Table 9.2. This table gives an idea of the finite-element grid needed. The typical element size ought to be around $\lambda/10$ to $\lambda/4$, depending on the element used (linear or quadratic). Discretisation problems increase when one considers processes with very short periods, such as earthquakes.

Table 9.2 TYPICAL WAVELENGTHS

	Exciting frequency	
	$T = 10\,\text{s}$	$T = 4\,\text{s}$
$E_{soil} = 3 \times 10^7\,\text{N/m}^2$	1000 m	400 m
$E_{soil} = 3 \times 10^5\,\text{N/m}^2$	100 m	40 m

It is now generally accepted that soil–foundation interaction is important and should be taken into account when calculating the response of the system. How important it is will depend on the particular soil and structure under consideration. Figure 9.11 shows a

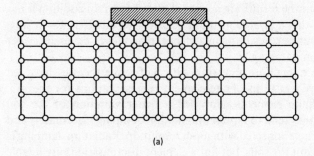

(a)

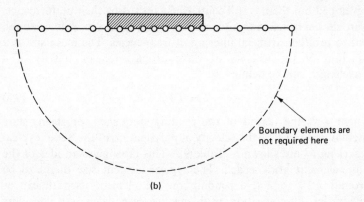

Boundary elements are
not required here

(b)

Figure 9.11   Representation of soil by (a) finite-element and (b) boundary-element mesh

structure idealised as a block on an elastic foundation that can only
have rocking movements. The period of the structure is assumed to be
$T_s = 3.5$ seconds and the wave frequency has a period of 12 seconds
(or $\omega = 0.52$ cycle/hour). To evaluate the importance of interaction,
one first determines the period of the system (assuming the structure
to be rigid) and then the period of the structure–soil system is found.
These periods were determined using the stiffness coefficients for
rocking presented in reference 7 (as shown in Table 9.3), which are
frequency dependent. The final soil–structure interaction period was
found using Dunkerley's laws, i.e. $T_{SF}^2 = T_S^2 + T_F^2$. Figure 9.12(a)
shows how the period varies when the soil modulus increases. For
hard soils the interaction effects become less important. Similarly if
the mass of the structure is reduced the interaction becomes less

Table 9.3 CONSTANTS FOR THE INFINITE SEMI-SPACE (FROM HALL AND KISSENPFENNIG[11] AND VELETSOS AND MEI[7])

Stiffness coefficients	Frequency independent	Frequency dependent
Horizontal translation	$k_x = \dfrac{32(1-v)GR}{7-8v}$	$\dfrac{32(1-v)GR}{7-8v}\left[1-0.05\omega R\sqrt{\left(\dfrac{\rho}{G}\right)}\right]$
Vertical translation	$k_z = \dfrac{4GR}{(1-v)}$	—
Rocking	$k_\theta = \dfrac{8GR^3}{3(1-v)}$	$\dfrac{8GR^3}{3(1-v)}\left[1-0.215\omega R\sqrt{\left(\dfrac{\rho}{G}\right)}\right]$
Torsion	$k_\phi = \dfrac{16GR^3}{3}$	—
Radiation damping coefficients	Frequency independent	Frequency dependent
Horizontal translation	$c_x = \dfrac{1.15\times\sqrt{[2(1-v)]\rho R^3}}{\sqrt{[m(7-8v)]}}$	$\dfrac{8G\omega R}{(2-v)}\sqrt{\left(\dfrac{\rho}{G}\right)}\left[0.67+0.02\omega R\sqrt{\left(\dfrac{\rho}{G}\right)}\right]$
Vertical translation	$c_z = \dfrac{0.85\times\sqrt{(\rho R^3)}}{\sqrt{[m(1-v)]}}$	—
Rocking	$c_\theta = \dfrac{0.3\times\sqrt{(2\rho R^5)}}{\sqrt{[3(1-v)I_\theta]}}\left(1+\dfrac{3(1-v)I_\theta}{8\rho R^5}\right)$	$\dfrac{0.375}{(1-v)}\omega^2\rho R^5$
Torsion	$c_\phi = \dfrac{0.5}{\left(1+\dfrac{2I_\phi}{\rho R^5}\right)}$	—

where ω = frequency of oscillation
G = shear modulus of the soil
R = radius of base
ρ = density of soil

v = Poisson's ratio for the soil
m = mass of foundation
I_θ = mass moment of inertia of foundation in rotation about a horizontal axis through the base
I_ϕ = mass moment of inertia of foundation in rotation about a vertical axis

316

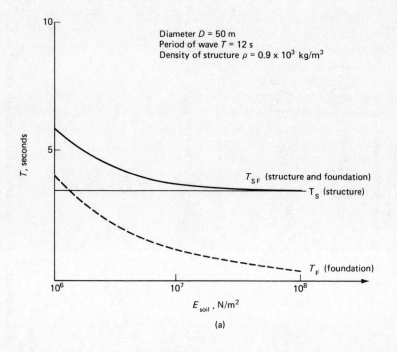

(a)

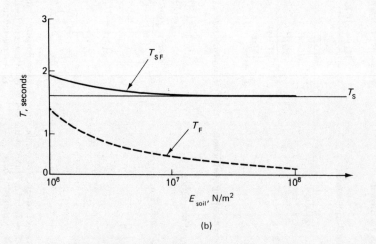

(b)

*Figure 9.12 Soil–structure interaction effect: (a) mass of structure = 2 × 10⁸ kg,
T_s = 3.5 seconds; (b) mass of structure = 2 × 10⁷ kg, T_s = 1.5 seconds*

marked; see Figure 9.12(b). This simple example serves to indicate the complexity of soil–structure interaction problems. The coefficients used here correspond to a rigid circular base; flexible foundations will behave somewhat differently.

Another important question is how important is the frequency dependence for the interaction coefficients. Figure 9.13 compares the case of using always the static coefficients $\omega = 0$ against two cases (for different ω) for which the variation with respect to frequency is taken into consideration. The graph shows that good results can be obtained in offshore structures using the static (i.e. constant frequencies) coefficients for low-frequency excitations or hard soils, but that for higher ω or soft foundations the dynamic coefficients are required.

Approximate procedures to determine the damping are generally

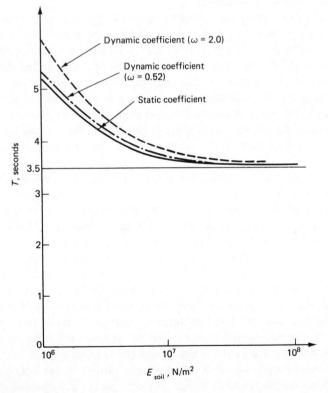

Figure 9.13 Static and dynamic interaction coefficients

used when working with modal analysis. Modal damping coefficients are expressed as weighted average of the soil and structural damping in proportion to the strain energy U. In such a way the weighted coefficient results[9]:

$$\beta = \frac{(\beta U)_{\text{soil}} + (\beta U)_{\text{structure}}}{U_{\text{structure}} + U_{\text{soil}}} \tag{9.9}$$

This simple procedure has two disadvantages. Firstly, soil damping is hysteretic in character whereas structural damping is viscous; in the above formulation viscous-type damping is assumed for the global system. Secondly, radiation damping is usually of a higher order of magnitude than structural and hydrodynamic damping. The validity of the above weighting procedure is limited to damping coefficients of similar orders of magnitude.

These disadvantages cast doubt on the validity of the weighted modal damping coefficients for soil–structure interaction problems. In our opinion the more adequate treatment is to use the eigen-modes in order to reduce the order of the system; this produces a coupled but small system of equations (the coupling is due to the non-diagonal damping matrix). One can work with the new matrix system without trying to force it to become an independent system of equations. This procedure has the main advantage of classical modal analysis but not the more important disadvantage, i.e. the need to diagonalise the damping matrix.

Methods such as semi-infinite space analyses give better results than finite elements but they are rather limited in their applications. Because of this, a new and more flexible method related to semi-infinite space solutions is now becoming widespread. This is the boundary element technique[10], which can be interpreted as a way of making a known solution (called the fundamental solution) satisfy the boundary conditions of the problem. If one considers a type of solution such as a point load in an infinite domain, obviously the conditions at infinity are well represented and radiation is adequately taken into account. The fundamental solution needs to be applied only on the boundaries instead of over all the domain. Furthermore, if the fundamental solution is not the one for an infinite space but for a *semi-infinite* one, the mesh becomes very simple as one only needs elements in the contact region between structure and soil. The advantages of using boundary elements in soil mechanics are obvious:

the number of unknowns reduces to those required at the finite boundary of the region under consideration.

The soil effects on piles are usually represented by a series of springs and dashpots. This method is often open to criticism, owing to the difficulty of finding the right coefficients, and a better approach is required. Boundary elements also provide a good method to determine the behaviour of piles or groups of piles (Figure 9.14). Layered soils can also be taken into consideration using boundary elements: the domain can be divided into different regions with a series of elements on the internal boundaries.

The main problem of soil–structure interaction analysis is the

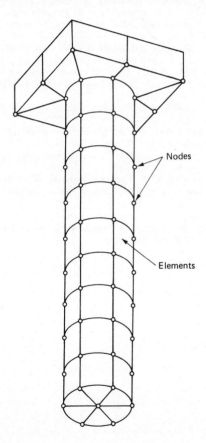

Nodes

Elements

Figure 9.14   Pile divided into boundary elements

representation of the material non-linearities. Hysteretic damping attempts to approximate the non-linear behaviour, but owing to the uncertainties involved in the process it is generally recommended to analyse the system with a range of soil properties and establish bounds in the results.

References

1. Ebert, M., *Some aspects of the analysis of offshore structures*, PhD thesis, Univ. Southampton (1977)
2. Vughts, J. H., and Kinra, R. K., 'Probabilistic analysis of fixed offshore structures', OTC 2608, Proc. Offshore Technology Conf., Houston (1976)
3. Kallaby, J., and Price, J. B., 'Evaluation of fatigue considerations in the design of framed offshore structures', OTC 2609, Proc. Offshore Technology Conf., Houston (1976)
4. Marshall, P. W., 'Dynamic and fatigue analysis using directional spectra', OTC 2537, Proc. Offshore Technology Conf., Houston (1976)
5. Nolte, K. G., and Hansford, J. E., 'Closed-form expression for determining the fatigue damage of structure due to ocean waves', OTC 2606, Proc. Offshore Technology Conf., Houston (1976)
6. Williams, A. K., and Rime, J. E., 'Fatigue analysis of steel offshore structures', *Proc. Instn Civil Engrs*, **60**, Part 1, 635–654 (Nov. 1976)
7. Veletsos, A. S., and Wei, Y. T., 'Lateral and rocking vibration of footings', *Proc. ASCE (Soil Mechanics Div.)*, **97** (Sep. 1971)
8. Luco, J. E., 'Independence function for a rigid foundation in a medium', *Nuclear Engng and Design*, **31**, 204–207 (1974)
9. Nataraja, R., and Kirk, C. L., 'Dynamic response of a gravity platform under random wave forces', OTC 2904, Proc. Offshore Technology Conf., Houston (1977)
10. Brebbia, C. A., and Walker, S., *Boundary elements*, Newnes–Butterworths (1979)
11. Hall, J. R., and Kissenpfennig, J. F., 'Special topics on soil–structure interaction', *Nuclear Engng and Design*, **38** (1976)

Index